Self-Assembly of Nanostructures and Patchy Nanoparticles

Edited by Shafigh Mehraeen

Published in London, United Kingdom

IntechOpen

Supporting open minds since 2005

Self-Assembly of Nanostructures and Patchy Nanoparticles
http://dx.doi.org/10.5772/intechopen.80196
Edited by Shafigh Mehraeen

Contributors
Jaroslav Ilnytskyi, Tianfeng Chen, Zusuang Xiong, Lanhai Lai, Thi Giang Le, Qian Chen, Ahyoung Kim, Lehan Yao, Falon Kalutantirige, Shan Zhou, Shafigh Mehraeen

Notice
Statements and opinions expressed in the chapters are these of the individual contributors and not necessarily those of the editors or publisher. No responsibility is accepted for the accuracy of information contained in the published chapters. The publisher assumes no responsibility for any damage or injury to persons or property arising out of the use of any materials, instructions, methods or ideas contained in the book.

First published in London, United Kingdom, 2020 by IntechOpen
IntechOpen is the global imprint of INTECHOPEN LIMITED, registered in England and Wales, registration number: 11086078, 5 Princes Gate Court, London, SW7 2QJ, United Kingdom
Printed in Croatia

British Library Cataloguing-in-Publication Data
A catalogue record for this book is available from the British Library

Additional hard and PDF copies can be obtained from orders@intechopen.com

Self-Assembly of Nanostructures and Patchy Nanoparticles
Edited by Shafigh Mehraeen
p. cm.
Print ISBN 978-1-78923-960-7
Online ISBN 978-1-78984-742-0
eBook (PDF) ISBN 978-1-78984-743-7

We are IntechOpen,
the world's leading publisher of
Open Access books
Built by scientists, for scientists

5,100+
Open access books available

126,000+
International authors and editors

145M+
Downloads

151
Countries delivered to

Our authors are among the

Top 1%
most cited scientists

12.2%
Contributors from top 500 universities

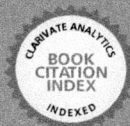

Interested in publishing with us?
Contact book.department@intechopen.com

Numbers displayed above are based on latest data collected.
For more information visit www.intechopen.com

Meet the editor

Shafigh Mehraeen is an Assistant Professor at the University of Illinois at Chicago. He received his M.Sc. and Ph.D. in Mechanical Engineering, both from Stanford University under the supervision of Andrew Spakowitz. For his Ph.D., he studied the impact of molecular elasticity on the behavior of semi-flexible polymers and protein self-assembly. As a postdoctoral scholar, he studied the impact of active layer morphology on bimolecular recombination losses in organic photovoltaics, and transition state theory under the supervision of Jean-Luc Bredas at Georgia Institute of Technology, and Jianshu Cao at MIT, respectively. He has published three books, more than 30 scientific papers, and served as a reviewer for major scientific journals. His current research focuses on applying molecular simulations, atomistic modeling, and density functional theory to address directed self-assembly of nanoparticles on templated surfaces, photochemistry of organic solar cells, and polymer and electrocatalyst design using machine learning and artificial intelligence.

Contents

Preface

Self-assembly of nanostructures has been identified as one of the important topics of nanoscience. Self-assembly, as a bottom-up approach, has the potential to lead future scientific research, including nanoelectronics, optoelectronics, spintronics, nanotechnology, nanobiotechnology, nanomanufacturing, drug delivery, materials science, robotics, and the like.

Recent works on self-assembled nanostructures have focused on the fundamental steps in crystal growth, chemical synthesis and self-organization of nanoparticles, quantum dots, 1D, 2D, and 3D nanostructures as well as their applications. These steps include nanocharacterization, lithographic techniques to apply nanopatterning, and mechanistic understanding of the self-assembly process from atomistic and molecular scale to the device scale. The main topics of interest encompass fabrication of (i) 0D nanostructures such as nanocrystals, quantum dots, and Q-bits; (ii) 1D nanostructures such as nanotubes, nanorods, and nanowires; (iii) 2D nanostructures such as ordered mesoporous oxides, graphene, nanomembranes, and silicene; and (iv) 3D nanostructures such as photonic crystals, bandgap materials, waveguides, and monolithic nanostructures. We believe that self-assembly will be a game changer in future nanotechnology and nanofabrication, and a technique for efficient, robust, and inexpensive manufacturing of unprecedented structures with defined functionalities.

This book is the result of concerted efforts to bring together studies in the broad field of research on self-assembly of nanostructures performed by reputable researchers in the field. The goal is to provide modern and advanced topics in self-assembly of nanostructures, nanomanufacturing techniques, and mechanistic understanding associated with the self-assembly process, not yet reflected by other books. As such, we focused on the physics of nanostructures at the nanoscale, including (1) large scale patterning formed by spontaneous structuring, (2) theoretical and experimental works to better understand the formation, progression, and organization of the self-assemblages, and (3) the new optical, electrical, mechanical, and medical properties of the assemblages. We hope that this book will serve as a guide for researchers, and inspire research ideas for potential future directions in the area of self-assembly of nanostructures and patchy nanoparticles.

This volume contains five chapters that cover the following directions. Chapter 1 - the Introductory chapter - briefly covers the latest developments in self-assembly of nanostructures, the core concepts, and active areas of research in the field. The remaining four chapters focus on different aspects of self-assembly of nanostructures. Every chapter provides an overview of the background for the subject matter, and description of the solution that was initially proposed by the authors.

Chapter 2 describes a computer simulation methodology to predict the assemblages of decorated nanoparticles, which are composed of a central core and a shell of ligands. The chapter dissects the impact of temperature, surface density, and type of ligands on the self-assembly morphology of the decorated nanoparticles. The chapter also highlights the application of chromophores in ligands, which enable photo-assisted self-assembly of the nanoparticles.

In Chapter 3, a novel approach to self-assembled nanorods on flat surfaces by means of molecular beam epitaxy is presented. It is shown that the nanocolumns have core-shell structures, while they grow perpendicular to the surface. A theoretical model is also presented to explain the driving forces, which were identified to be the diffusion of constituents in the lateral direction and surfactant effect in the longitudinal direction.

Chapter 4 gives a detailed introduction to the self-assembly of patchy particles and the principles of their synthesis. The chapter highlights the importance of surface anisotropy in self-assembly of nanoparticles such as Janus particles, virus capsids *in vivo*, and globular proteins with patches, which function as the recognition sites. The surface anisotropy gives rise to highly directional interactions, and facilitates nanofabrication of ordered nanostructures by means of self-assembly.

Chapter 5 addresses one of the most important applications of self-assembled nanostructures in cancer treatment. The authors discuss the rational design of various metalopolymers, which is a combination of organic polymers and metal centers, and control of molecular interactions towards synthesis of self-assembled functional supramolecular metalopolymers.

In summary, we present a wide range of topics in the book, which can be addressed to a broad audience, from researchers in the academia to practitioners in the industry sector, and those who are interested in self-assembly of nanostructures and patchy nanoparticles, and their applications. This book is the result of several collaborating groups. I gratefully acknowledge all the authors who contributed to this book.

Shafigh Mehraeen
Department of Chemical Engineering,
University of Illinois at Chicago,
USA

Chapter 1

Introductory Chapter: Self-Assembly of Nanostructures

Shafigh Mehraeen

1. Introduction

Self-assembly is a process by which building blocks are put together in a parallel fashion without external stimuli. The structures and patterns that emerge from the self-assembly can be small or large, depending on the interaction forces between the building blocks and their kinetic pathways during the self-assembly. Self-assembly dates back to 400 BC, when Democritus, a Greek philosopher, envisioned that the earth and solar system might have evolved from organization of atomistic components [1]. Later, Descartes, a French philosopher, postulated that the universe has self-structured out of chaos via organization of small components to large assemblages according to natural laws of physics [1]. In 1935, Langmuir and Blodgett developed a method to form a closely packed monolayer of amphiphilic molecules on solid and liquid surfaces [2]. In 1946, Bigelow observed the assembly of a monolayer of long alkylamine chains on a solid surface. Although these studies did not use the term "self-assembly", they were indirectly explaining the self-assembly process. The self-assembly was initially used in Self-assembled monolayers (SAMs), which were discovered in 1983, when Nuzzo et al. developed well-ordered monolayers of alkanethiolate molecules chemosorbed on gold surfaces [3–5].

Aforementioned ideas, observations, and discovery have led to the current thinking of fabricating nanostructures via self-assembly, which is spontaneous organization of ordered structures from disordered phase of nanoscale constituents in the absence of external stimuli. Self-assembly manifests a bottom-up approach in nanofabrication, and enables an efficient way to create functional materials from nanoscale building blocks. Current nanotechnology utilizes two approaches for nanomanufacturing: (i) top-down approach, where materials are manipulated at the macroscopic scales in serial steps to fabricate functional construct at the nanoscale, and (ii) bottom-up approach, where materials are constructed by assembly of nanoscale building blocks to form functional construct at macroscopic scales.

Controlling the size, shape, and surface properties of the building blocks determines how ordered the organization of constructs in the self-assembly will be. As such, synthesizing the nanoscale building blocks with desired shape, size, and properties is the primary goal in the self-assembly. In addition, with specific properties of the building blocks, e.g. charge, binding affinity, and hydrophobicity/hydrophilicity, one will be able to control the interaction forces between the building blocks, and their kinetic pathways during the self-assembly. These interaction forces can span multiple length scales. In the absence of external stimuli, the self-assembled configuration is settled by overall energy minimization of the system. This self-assembled structure will be static, and correspond to a local minimum energy configuration, which may be kinetically trapped. However, in the presence of external stimuli, the self-assembled configuration will correspond to a local energetic minimum, which depends on the influx of energy from the external stimuli. As soon as the external stimuli cease to exist, the energetic minimum will disappear, and the assembled configuration will dismantle.

2. "Hierarchical" and "directed" self-assembly

Self-assembly process can be done in various ways among which "hierarchical" and "directed" self-assembly are very common in nanofabrication. In hierarchical self-assembly, structuration of the building blocks occurs over multiple length scales. Initially, at the smallest length scale, self-assembly of the original building blocks takes place. The assemblages that grow out of the first level of assembly form the building blocks for the second stage. This hierarchy of self-assembled building blocks continues at multiple levels, which span multiple length scales. An example of hierarchical self-assembly is the mesostructure of Siloxane-organic hybrid films with ordered macropores [6]. These hybrid films have spanned three length scales during self-assembly, (i) microscopic length scale at which assembly of tetrahedral silica blocks in the presence of polystyrene particle template takes place, (ii) mesoscopic length scale at which ordered mesoscopic pores form, and (iii) macroscopic length scale at which macroscopic air voids with silica intercalated network is assembled.

In contrast to hierarchical self-assembly, in the directed self-assembly an external stimulus is used to drive the assembly in a desired direction. One way to direct the self-assembly is to provide a lithographic template, and use the template to guide colloidal particles or nanoparticles in a solution to self-assemble on the template surface [7]. One example of directed self-assembly is the dip-pen nanolithography, whereby AFM tip and water meniscus is utilized to direct chemical reagents to form a SAM on nanoscopic areas of the target surface [8]. Directed self-assembly can further be divided to static and dynamic self-assembly. In static self-assembly, the external forces drive the system to an equilibrium state. The system dissipates energy as it moves towards a static equilibrium. At this equilibrium, the self-assembled structure will retain its configuration there after even if the external stimuli are removed; hence called static self-assembly. In contrast, dynamic self-assembly generates stable assemblages, which may not be at thermodynamic equilibrium. These stable configurations persist as long as the external stimuli are present and the system dissipates energy [9]. These configurations can also be manipulated by rectifying the external forces. As soon as external forces are removed, the system will disassemble, and return to a thermodynamic equilibrium. An example of dynamic self-assembly is the self-organization of rotating magnetic disks floating at liquid-air interface in the presence of an external magnetic force, which attracts the disks towards the axis of rotation. As disks move at the interface, they create vortices, which repel the disks away from another. The balance between the attractive and repulsive forces gives rise to the dynamic self-assembled structure of the magnetic disks [10].

3. Conclusion

Aforementioned applications are only a short list among many others, which suggest that self-assembly is a promising bottom-up approach that can potentially lead to a scalable and efficient nanofabrication technique. Nanofabrication of most of complex three-dimensional structures [11], functional materials [12], nanocrystal superlattices that would look like nanorods, and inverse opals [6], or manufacturing nanowires with diameters less than 2 nm [13] have been reported by means of self-assembly; however, fabrication of these structures will pose a challenge, and be costly by top-down approaches. On the other hand, despite the advantages of self-assembly, and its promising potentials in nanofabrication, there are still challenges to utilize it for certain applications. These challenging applications include but not limited to:

(1) Designing semiconductors for inexpensive light harvesting systems with high power conversion efficiency, and enhanced thermal stability, (2) fabricating superlattices of quantum dot semiconductors with highly efficient refrigeration and power generation cycles for thermoelectric materials, (3) synthesizing solid-state light weight nanomaterials for lithium batteries with high energy density, and fast charge-discharge cycles, (4) positioning of nanoparticles with single particle resolution and sub-10 nm precision for nanoscale electronics, (5) synthesizing nanomaterials with highly selective electrocatalytic activities, (6) developing nanomaterials for carbon sequestration, and efficient conversion, (7) developing porous structures for fast hydrogen storage at atmospheric pressure, room temperature, and reversible usage, (8) developing a low cost, portable, rapid, and sensitive disease diagnostic sensors, (9) designing new drug delivery agents for treatment of targeted cancer cells, and (10) fabricating nanomotors with speed and direction controllability for payload delivery in biomedical applications.

Author details

Shafigh Mehraeen
Chemical Engineering, University of Illinois at Chicago, United States of America

*Address all correspondence to: tranzabi@uic.edu

IntechOpen

References

[1] B. Russell, A history of western philosophy. 1972.

[2] Blodgett KB. *Films built by depositing successive monomolecular layers on a solid surface*. Journal of the American Chemical Society. 1935;**57**:1007-1022. DOI: 10.1021/ja01309a011

[3] Nuzzo RG, Allara DL. *Adsorption of bifunctional organic disulfides on gold surfaces*. Journal of the American Chemical Society. 1983;**105**:4481-4483. DOI: 10.1021/ja00351a063

[4] Laibinis PE, Whitesides GM, Allara DL, Tao YT, Parikh AN, Nuzzo RG. *Comparison of the structures and wetting properties of self-assembled monolayers of n-alkanethiols on the coinage metal surfaces, copper, silver, and gold*. Journal of the American Chemical Society. 1991;**113**:7152-7167. DOI: 10.1021/ja00019a011

[5] Troughton EB, Bain CD, Whitesides GM, Nuzzo RG, Allara DL, Porter MD. *Monolayer films prepared by the spontaneous self-assembly of symmetrical and unsymmetrical dialkyl sulfides from solution onto gold substrates: Structure, properties, and reactivity of constituent functional groups*. Langmuir. 1988;**4**:365-385. DOI: 10.1021/la00080a021

[6] Sakurai M, Shimojima A, Heishi M, Kuroda K. *Preparation of Mesostructured siloxane–organic hybrid films with ordered macropores by templated self-assembly*. Langmuir. 2007;**23**:10788-10792. DOI: 10.1021/la701590x

[7] Aizenberg J, Braun PV, Wiltzius P. *Patterned colloidal deposition controlled by electrostatic and capillary forces*. Physical Review Letters. 2000;**84**:2997-3000. DOI: 10.1103/PhysRevLett.84.2997

[8] Ginger DS, Zhang H, Mirkin CA. *The evolution of dip-pen nanolithography*. Angewandte Chemie International Edition. 2004;**43**:30-45. DOI: 10.1002/anie.200300608

[9] Fialkowski M, Bishop KJM, Klajn R, Smoukov SK, Campbell CJ, Grzybowski BA. *Principles and implementations of dissipative (dynamic) self-assembly*. The Journal of Physical Chemistry B. 2006;**110**:2482-2496. DOI: 10.1021/jp054153q

[10] Grzybowski BA, Campbell CJ. *Complexity and dynamic self-assembly*. Chemical Engineering Science. 2004;**59**:1667-1676 https://doi.org/10.1016/j.ces.2004.01.023

[11] Ghadimi A, Cademartiri L, Kamp U, Ozin GA. *Plasma within templates: Molding flexible nanocrystal solids into multifunctional architectures*. Nano Letters. 2007;**7**:3864-3868. DOI: 10.1021/nl072026v

[12] Lotsch BV, Knobbe CB, Ozin GA. *A step towards optically encoded silver release in 1D photonic crystals*. Small. 2009;**5**:1498-1503. DOI: 10.1002/smll.200801925

[13] R. S. Friedman, M. C. McAlpine, D. S. Ricketts, D. Ham, C. M. Lieber, "High-speed integrated nanowire circuits," Nature 434, 1085-1085, 2005. 10.1038/4341085a

Self-Assembly of Nanoparticles Decorated by Liquid Crystalline Groups: Computer Simulations

Jaroslav Ilnytskyi

Abstract

We present the results of the computer simulations for the self-assembly of decorated nanoparticles. The models are rather generic and comprise a central core and a shell of ligands containing terminal liquid crystalline group, including the case of the azobenzene chromophores. The simulations are performed using the coarse-grained molecular dynamics with the effective soft-core interparticle interaction potentials obtained from the atomistic simulations. The discussion is centred around the set of the self-assembled morphologies in a melt of 100–200 of such decorated nanoparticles obtained upon the change of the temperature, surface density of ligands, the type of the terminal group attachment, as well as the prediction of the possibility of photo-assisted self-assembly of the nanoparticles decorated by the azobenzene chromophores.

Keywords: self-assembly, nanoparticles, molecular dynamics, azobenzene

1. Introduction

Nanoparticles gained extended research and technological focus due to their unique optical, electronic, magnetic and chemical properties [1]. Applications include: medical diagnostics, drug delivery, cancer therapy, nanoelectronics and information storage, sensors, (photo)catalysis, surface coatings [2]. Self-assembly of nanoparticles are in a core of many advanced materials developments coining the term 'nanoarchitectonics' [3]. Self-assembly process becomes more controllable by decoration of nanoparticles with functional ligands. Good example is decorated nanoparticles (DNP) coated by the liquid crystalline (LC) ligands [4–9] that are considered in this chapter.

The structure of the self-assembled morphology depends on: (i) the details of molecular architecture, and (ii) external conditions. The group (i) includes the type of the core nanoparticle (metal/non-metal, magnetic/non-magnetic, etc.); grafting density, length, flexibility and chemical details of ligands; number and type of functionalisation groups, etc. The group (ii) includes the temperature, pressure/density, the presence of surfaces and external fields. Prediction of symmetry/structure/properties of the self-assembled morphology, especially by means of theoretical methods, is quite problematic.

Computer simulations are of great aid here, as these incorporate the relevant details of molecular architecture and tackle statistical behaviour of many-particle systems under specified external conditions. This approach faces a difficulty in terms of a wide span of the time- and length-scales typical for the supramolecular self-assembly. In particular, if more chemistry-specific details are taken into account—then relatively small system sizes and short simulation times can be covered. A good compromise can be achieved by using elements of a multiscale approach [10, 11], which builds a coarse-grained model based on the simulation data of a more chemically-detailed model. A coarse-grained model is of rather generic type capturing essential physical details of the atomistic system and allowing to reach required time- and length-scales of a self-assembly [12, 13]. Due to inevitable loss of specific chemical details, the comparison with particular chemical realisations is performed on a high level only—via the structure of observed morphologies and via the temperature- or density-driven phase transitions between them.

We cover the details of a coarse-grained modelling and self-assembly of DNPs containing LC groups (including the case of azobenzene chromophores). Computer simulations are performed via the molecular dynamics simulation combined with stochastic photoisomerisation events (in the case of azobenzene chromophores). As the result, this type of modelling enables to consider the role of grafting density and type of LC group attachment, as well as the role of the temperature and external fields (including illumination) on the process of self-assembly. Section 2 contains modelling and simulation details, in Section 3 we consider temperature related effects of the self-assembly, Section 4 covers the role of the details of molecular architecture, in Section 5 we cover the photo-aided self-assembly of DNPs containing azobenzene chromophores.

2. Modelling and simulation details

To study self-assembly of DNPs we use coarse-grained modelling, where relevant groups of atoms are replaced by single beads that interact via soft-core potentials. The model DNP is represented schematically in left frame of **Figure 1**. It is built from a central core and N_{ch} ligands that are free to surf on its surface. Each ligand is terminated by a LC group. The model equally may represent a generation three carbosilane dendrimer [14, 15].

The model contains spherical beads (designated thereafter as 'sp') and the spherocylinder ('sc') ones that mimic LC groups. The sphere-sphere interaction is soft repulsive of quadratic form

$$V_{ij}^{sp-sp} = \begin{cases} U_{max}^{sp-sp}\left(1 - r_{ij}^*\right)^2, & r_{ij}^* < 1 \\ 0, & r_{ij}^* \geq 1, \end{cases} \tag{1}$$

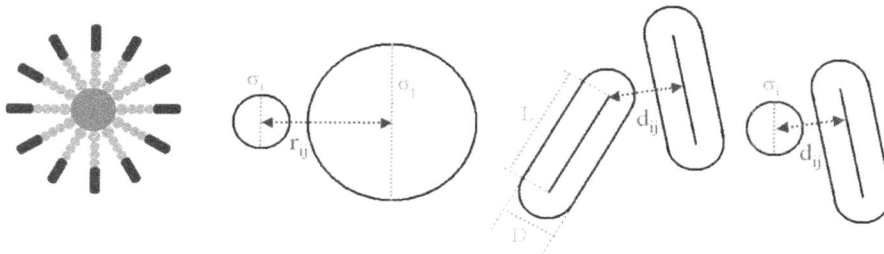

Figure 1.
Model DNP (left frame). Definition of sizes and distances for pairs of interacting beads (right frame).

where $r_{ij}^* = r_{ij}/\sigma_{ij}$ is scaled distance between centres of ith and jth sphere and mixing rules $\sigma_{ij} = (\sigma_i + \sigma_j)/2$ are implied (see right frame of **Figure 1**). $U_{\max}^{\text{sp-sp}}$ is the same for all interacting spheres.

Spherocylinders are of breadth D and of elongation L/D and interact via the soft anisotropic potential of Lintuvuori and Wilson [16]:

$$V_{ij}^{\text{sc-sc}} = \begin{cases} U_{\max}^{\text{sc-sc}}\left(1 - d_{ij}^*\right)^2, & d_{ij}^* < 1 \\ U_{\max}^{\text{sc-sc}}\left(1 - d_{ij}^*\right)^2 - U_{\text{attr}}^*\left(\hat{r}_{ij}, \hat{e}_i, \hat{e}_j\right)\left(1 - d_{ij}^*\right)^4 + \epsilon^*, & 1 \leq d_{ij}^* < d_c^* \\ 0, & d_{ij}^* > d_c^*, \end{cases} \quad (2)$$

where $d_{ij}^* = d_{ij}/D$ is the dimensionless nearest distance between the cores of spherocylinders (see right frame of **Figure 1**), d_c^* is the orientation dependent cutoff. Attractive interaction has the form:

$$U_{\text{attr}}^*\left(\hat{r}_{ij}, \hat{e}_i, \hat{e}_j\right) = U_{\text{attr}}^* - \left[5\epsilon_1 P_2\left(\hat{e}_i \cdot \hat{e}_j\right) + 5\epsilon_2\left(P_2\left(\hat{r}_{ij} \cdot \hat{e}_i\right) + P_2\left(\hat{r}_{ij} \cdot \hat{e}_j\right)\right)\right] \quad (3)$$

and depend on orientations $\hat{e}_i$ and $\hat{e}_j$ of the long axes of spherocylinders and the unit vector $\hat{r}_{ij}$ that connects their centres [16]. $P_2(x) = 1/2(3x^2 - 1)$ is the second Legendre polynomial, energy parameters U_{attr}^*, ϵ_1 and ϵ_2 are given below. The inclusion of the attractive contribution in Eq. (2) shifts the region for the LC stability towards smaller elongations $L/D \sim 3$ compared to the case of purely repulsive interactions, where, typically, $L/D \sim 6 - 8$ [14].

Mixed nonbonded interactions are evaluated in a similar way to Eq. (1)

$$V_{ij}^{\text{sp-sc}} = \begin{cases} U_{\max}^{\text{sp-sc}}\left(1 - d_{ij}^*\right)^2, & d_{ij}^* < 1 \\ 0, & d_{ij}^* \geq 1, \end{cases} \quad (4)$$

where $d_{ij}^* = d_{ij}/\sigma_{ij}$ is a dimensionless distance between the centre of the ith sphere and the core of the jth spherocylinder (see right frame of **Figure 1**), with the scaling factor $\sigma_{ij} = (\sigma_i + D)/2$.

Intramolecular interactions for the model include bond and angle interactions terms

$$V_{\text{bonded}} = \sum_{i=1}^{N_b} k_b\left(l_i - l_0^k\right)^2 + \sum_{i=1}^{N_a} k_a\left(\theta_{ijk} - \theta_0\right)^2, \quad (5)$$

where l_i is the instantaneous bond length, $\left\{l_0^k\right\}$ is the set of effective bond lengths. Ligands stiffness is adjustable via the magnitude of k_a, and θ_0 is set equal to π.

The force-field parameters are based on the coarse-graining of the LC dendrimer first performed and described in detail in Ref. [14]. Bulk behaviour of such macro-molecules are very similar to that of the DNPs [17]. The soft-core diameters of spherical beads are: $\sigma = 21.37$, 6.23 and 4.59 Å for the core, first ligand bead and the following ligans beads, respectively. Their respective masses: $62.44 \cdot 10^{-25}$, $2.20 \cdot 10^{-25}$ and $0.70 \cdot 10^{-25}$ kg. LC beads dimension are $D = 3.74$ Å, $L/D = 3$, their mass is $3.94 \cdot 10^{-25}$ kg and the moment of inertia: $6.00 \cdot 10^{-24}$ kg. The energy parameters $U_{\max}^{\text{sp-sc}}$, $U_{\max}^{\text{sp-sc}}$ and $U_{\max}^{\text{sp-sc}}$ are all equal to $70 \cdot 10^{-20}$ J. LC-LC specific energy param-eters are $U_{\text{attr}}^* = 1500 \cdot 10^{-20}$, $\epsilon_1 = 120 \cdot 10^{-20}$ and $\epsilon_2 = -120 \cdot 10^{-20}$ J. The bond lengths are: 14.9, 3.60 and 3.62 Å between the core-first spacer bead, first-second spacer bead and between following spacer beads, respectively. The spherocylinder is attached to the centre of its nearest spherical cap with the bond length of 2.98 Å. The bond interaction spring constant is $50 \cdot 10^{-20}$ J for all bonds. The pseudo-valent angle spring constant is $20 \cdot 10^{-20}$ J.

The simulations are carried out with the GBMOLDD program extended to the case of the $NP_xP_yP_zT$ ensemble [18]. We use a single Nóse-Hoover thermostat in most cases for both translational and rotational degrees of freedom. In quenching or rapid heating runs the velocity rescaling was used instead. The timestep $\Delta t \sim 20$ fs was used for velocity rescaling runs, whereas smaller timesteps $\Delta t \sim 10 - 15$ fs was required for the runs with the thermostat. To control pressure three barostats are used [18].

3. Temperature driven morphology changes

In this section we consider temperature driven transitions between ordered and disordered morphologies observed in the DNPs melt. The experimental evidence for such transitions are found in Refs. [4–9, 17] and indicate a close relation between the average DNPs shape and the type of the ordered morphology. Therefore, we attempt to steer the self-assembly towards particular morphology by influencing the DNPs shape. To do so we impose the orientation field of a given symmetry, which acts on the LC beads. It is introduced via the following energy term

$$U_{\text{rot}} = -F(\mathbf{e}_i \cdot \mathbf{i})^2, \tag{6}$$

where $\mathbf{e}_i$ is the orientation of ith LC bead, $\mathbf{i}$ is the direction of the field, whereas field strength F provides either uniaxial ($F>0$) or planar ($F<0$) preferred orientation of the LC beads. In turn, in these cases we expect, respectively, rod-like or disc-like conformations of DNPs.

The case of the uniaxial field is considered first. Applied to the isotropic morphology of 100 DNPs in bulk using the $NP_xP_yP_zT$ ensemble at $P = 50$ atm and $T = 520$ K, the field of the strength $F = 2 \cdot 10^{-20}$ J and with $\mathbf{i}$ collinear to the Z-axis, it induces formation of a monodomain smectic A (Sm$_A$) morphology. It was studied then with the field switched off and the melt equilibrated for $20 - 40$ ns at selected temperatures within a range of $T \in [350 \text{ K}, 650 \text{ K}]$. Its appearance is shown in **Figure 2** indicating a lamellar structure with alternating layers of cores, ligand chains and LC beads (shown in left frame), in agreement with the experimental studies [5, 17]. The two-dimensional arrangement within layers is shown in the right frame.

We examined spatial distribution of the DNPs cores next. The form of their radial distribution function $g(r)$ is similar within a range of $T = 350 - 470$ K indicating two maxima: one at $r \sim 27$ Å for the short-range order of cores within each layer, and another at $r \sim 65$ Å, related to the interlayer distance (see, **Figure 3**, left frame). These characteristic distances are model dependent and can be examined by evaluting pair distribution functions $g_z(r)$ and $g_{xy}(r)$. The former yields an interlayer distance at ~ 59 Å (**Figure 3**, right frame). The latter indicates some degree of local (but not long-ranged) positional order within the layers (**Figure 3**, middle frame), characteristic of a 2D liquid. Therefore the phase is identified as the smectic A (Sm$_A$) [5, 17].

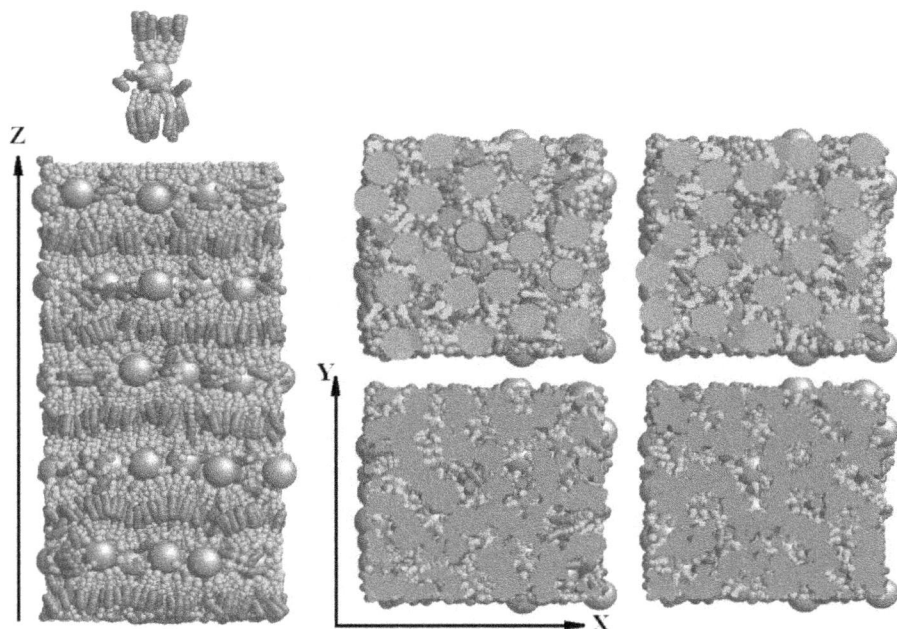

Figure 2.
Snapshots of the Sm_A phase at $T = 470\,K$ including typical rod-like molecular conformation.

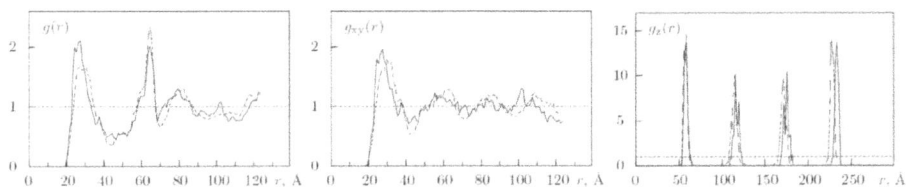

Figure 3.
Radial distribution functions $g(r)$ (left frame), $g_{xy}(r)$ (middle frame) and $g_z(r)$ (right frame) for the DNPs cores in the Sm_A phase at $T = 350\,K$ (solid lines) and at $T = 470\,K$ (dashed lines).

Application of a field with a planar symmetry ($F < 0$) along the Z axis, leads to the formation of a columnar (Col) morphology. The defect-free monodomain sample is achieved by using the field of a moderate magnitude, $F = -2 \cdot 10^{-20}\,J$ applied to the isotropic melt at $P = 50$ atm and $T \sim 520$ K. It was subsequently studied in a series of runs at the temperatures in the range $T \in [350\,K, 650\,K]$ and is shown at $T = 490$ K in **Figure 4**. It displays disc-like DNPs stacked into columns, whereas the columns theirselves are arranged hexagonally, in accordance with the experimental observations [4].

The structure of Col phase is analysed via the pair distribution functions, shown in **Figure 5**. The hexagonal arrangement of columns is clearly indicated in the form of $g_{xy}(r)$, while the peaks in $g_z(r)$ allow the distance between dendritic cores within a column to be estimated as $\sim 18\,\text{Å}$ for the Col phase at $T = 350$ K and ~ 20 Å at $T = 490$ K.

Upon heating up, both ordered morphologies, Sm_A and Col, undergo transitions to the spatially disordered phase as monitored via the order parameter S_N of

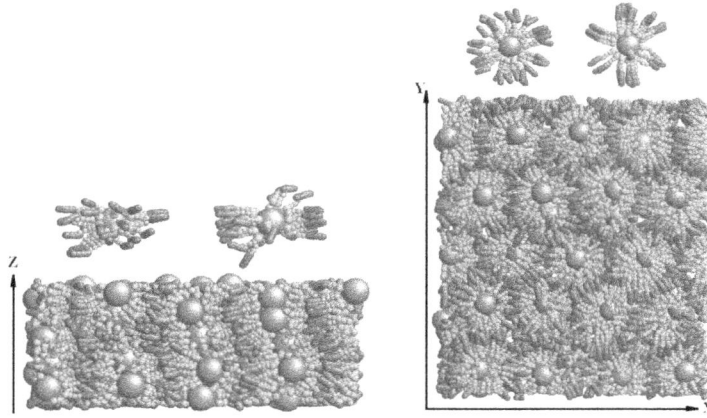

Figure 4.
Snapshots of the Col phase at $T = 490\,K$ including typical disc-like molecular conformation.

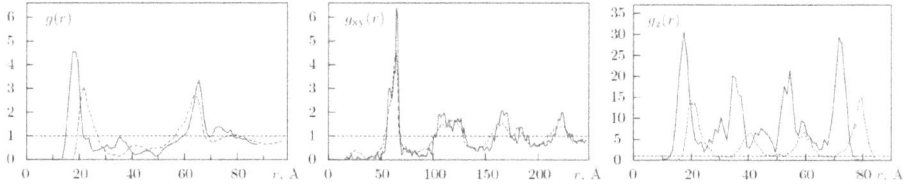

Figure 5.
Radial distribution functions $g(r)$ (left frame), $g_{xy}(r)$ (middle frame) and $g_z(r)$ (right frame) for the DNPs cores in the Col phase at $T = 350\,K$ (solid lines) and at $T = 470\,K$ (dashed lines).

mesogens with respect to the axis **i** (nematic director for the Sm$_A$ and the columns vector for the Col symmetry)

$$S_N = \langle P_2(\mathbf{e}_i \cdot \mathbf{i}) \rangle, \tag{7}$$

where $P_2(x)$ is the second Legendre polynomial and averaging is performed on all mesogens in a melt. As far as in this section the field is always collinear to the Z-axis, the notation S_z is also used. Shape asymmetry of DNPs, a, is obtained from the components of the gyration tensor $G_{\alpha\beta}$

$$G_{\alpha\beta} = \frac{1}{N} \sum_{i=1}^{N} (r_{i,\alpha} - R_\alpha)(r_{i,\beta} - R_\beta), \quad a = \frac{1}{R_g^2}\left[G_{zz} - \frac{G_{xx} + G_{yy}}{2} \right], \tag{8}$$

where α, β stay for the Cartesian axes, $r_{i,\alpha}$ and R_α are the coordinates of the individual beads and of the DNP's centre of mass, respectively, $R_g^2 = \sum_{\alpha\alpha} G_{\alpha\alpha}$ is the squared radius of gyration. Each spherocylinder bead is replaced by a line of four centres. The shape anisotropy a is positive for the prolate shape and negative for the oblate one.

S_z, a and the system density ρ are all shown as the functions of the temperature in **Figure 6** for both Sm$_A$ (left frame) and Col (right frame) morphologies. At lower temperatures both S_z and a are non-zero, indicating the Sm$_A$ and Col phases and quantifying the amount of orientation order and DNPs shape asymmetry at each temperature. Transition to the disordered isotropic morphology occurs at about $T = 490 - 500\,K$, where both S_z and a simultaneously turn to zero.

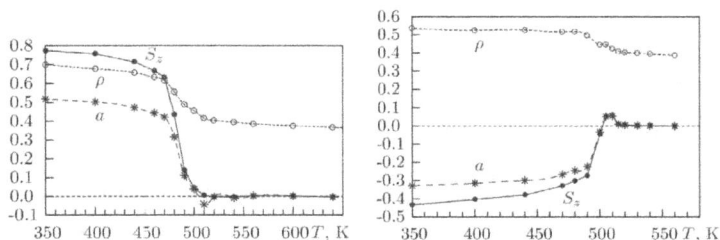

Figure 6.
Evolution of the density, shape asymmetry and the order parameter at the Sm_A-I and Col-I phase transitions.

The simulations reproduce a strong link between the shape of a DNPs and the type of bulk morphology, as previously observed experimentally [4, 17]. In particular, rod shapes are compatible with the Sm_A symmetry (**Figure 2**), discotic shapes are compatible with Col symmetry (**Figure 4**) and an spherical molecular shape is found in an isotropic state. This confirms the validity of the present model, which catches main features of the DNPs bulk assembly. The prolate-to-isotropic and oblate-to-isotropic shape transition occur simultaneously with vanishing the respective order parameter, as shown in **Figure 6**. The DNPs demonstrate shape bistability leading to the possibility to observe different symmetries at the same thermodynamics condition. As will be shown in the following section, this feature is dependent of the density of ligands defined via N_{ch}.

4. The role of the details of the molecular architecture

4.1 Variation of a grafting density

In Section 3 the number of ligands in DNPs was kept fixed at $N_{ch} = 32$. The decoration density, however, can change the shape of the supermolecule leading to different types of self-organised structure [4, 6, 17]. On the first sight, the most favourable conformation at any N_{ch} could be estimated from space-filling considerations and then the type of self-assembled morphology could be predicted. This was attempted in [19] but showed that the temperature effects also plays a crucial role and the symmetry of the self-assembled morphology at each conditions is the result of a delicate balance between the enthalpic and entropic contributions to the free energy.

Based on experimental findings [5, 8, 9, 17, 20], we expect to observe the sequence of Sm_A, Col and cubic morphologies upon the increase of N_{ch}. For each N_{ch}, we performed aided self-assembly runs of duration 20 ns at $T = 520K$ with the timestep of 20 fs in the $NP_xP_yP_zT$ ensemble. Both uniaxial and planar fields were used. Then, the field was switched off and the system was equilibrated at $T = 450$ K, about 50 K below the transition. Besides these, the spontaneous self-assembly runs were performed, too. In the latter, the temperature of the melt was reduced linearly from $T = 500K$ down to 450 K during first 20 ns (cooling rate is 2.5 K/ns), followed by another run for 20 ns at fixed $T = 450K$. As the result, relatively defect-free smectic layers are found for the cases of $N_{ch} = 12$ and $N_{ch} = 20$ only, whereas at $N_{ch} \leq 24$ we obtained polydomain layered structures with either one type of domains (of smectic or discotic type) or a mixture of both. In the field-aided self-assembly the external field (6) was employed, similarly as described in

Section 3. Both runs with uniaxial ($F>0$) and planar ($F<0$) fields were undertaken for each case of the number of ligands N_{ch} being considered.

Due to a relatively small system size, firm identification of smectic and discotic domains turned to be quite problematic, the visual inspection indicates domains of just a few DNPs. Therefore, we opted to analyse the distribution of the DNPs shape asymmetries instead, assuming that each rod-shaped DNP is a part of a smectic domain and a disc-shaped one—of the discotic one, as demonstrated earlier in Section 1. To distinguish between two shapes, we introduce molecular 'roddicity' a_r (always positive) and molecular 'discoticity' a_d (always negative):

$$a_r = \frac{1}{R_g^2}\left[\lambda_{\max} - \frac{\lambda_{\mathrm{med}} + \lambda_{\min}}{2}\right], \qquad a_d = \frac{1}{R_g^2}\left[\lambda_{\min} - \frac{\lambda_{\mathrm{med}} + \lambda_{\max}}{2}\right], \qquad (9)$$

for each DNP. Here $\lambda_{\max}$, λ_{med} and $\lambda_{\min}$ are the maximum, medium and minimum eigenvalues of the gyration tensor (8), respectively. If, for given DNP, $|a_r|>|a_d|$, then it is classified as a rod with the shape anisotropy $a' = a_r$, otherwise— as a disc with $a' = a_d$. At each time instance, the DNP melt splits into rods and discs subsystems, with their fractions termed as f_r and f_d, respectively.

The histograms for the distribution $p(a')$ are built over all the DNPs in the system averaged over time trajectory and are shown in **Figure 7** for selected values of N_{ch}. Both the cases of unaided (left frame) and field-aided self-assembly (middle and right frames) are displayed. In the case of spontaneous self-assembly, rods and discs always coexist and the distributions of their shape asymmetry are relatively broad. With the increase of N_{ch}, two maxima gradually merge into a spherulitic shape from both sides of $a' = 0$ (at about $N_{ch} = 64$ and higher). The histograms at the field-aided self-assembly are much narrower. In the case of uniaxial field, the discotic conformations are completely eliminated (except the case of $N_{ch} = 48$ where smectic phase is not observed any more), as these are incompatible with the 1D symmetry of the aiding field. In the case of planar filed, which has a 2D symmetry, the rod-like conformations are not eliminated and do appear within XY plane, and are, in fact, the dominant ones at smaller values of N_{ch}. With the increase of N_{ch} above 24 the disc-like conformations are dominating.

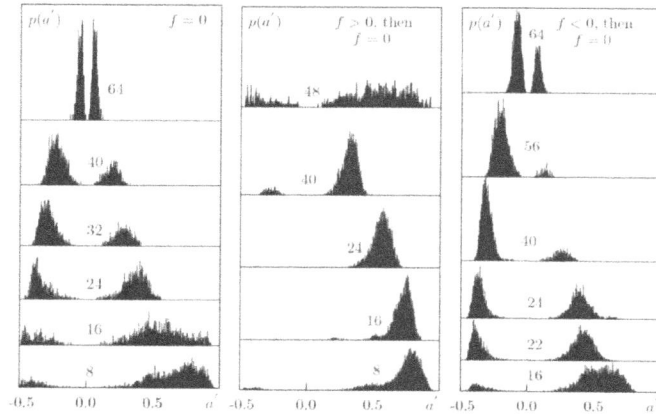

Figure 7.
Histograms for the distributions of shape anisotropy $p(a')$ at spontaneous self-assembly (left image), uniaxial field aided self-assembly (middle image) and planar field aided self-assembly (right image).

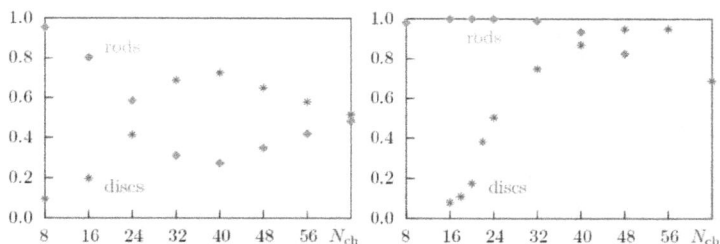

Figure 8.
Fraction of rods and discs for unaided self-assembly (left frame, $f_r + f_d = 1$). The same properties are shown on the right but fraction of rods is shown for uniaxial field aided runs and fraction of discs—for planar field aided runs.

Resulting fractions of rods and discs, f_r and f_d, are shown in **Figure 8** as functions of N_{ch} for both types of runs. Left frame (unaided self-assembly case) indicates the broad region of the rod-disc coexistence at intermediate values of N_{ch}. At $N_{ch} = 64$ the system approaches symmetric case with both conformations transforming into a spherulitic shape. The right frame contains data for f_r for uniaxial field aided self-assembly and data for f_d for planar field aided self-assembly, therefore, $f_r + f_d \neq 1$ as both are obtained for different conditions. The shapes of both curves are much steeper as compared to the left frame plot indicating the possibility to control the molecular conformation by means of initial field of appropriate symmetry. Therefore, there is some interval of grafting density, where DNPs exhibit rod-disc shape bistability and, hence, the Sm_A-Col morphology bistability is observed in the melt. Spontaneous self-assembly yields the polydomain structure with both Sm_A and Col fragments. Sm_A or Col morphology can be made dominant or, at least, enhanced by an external field of appropriate symmetry [19]. One of such cases is discussed in detail in Section 5. These results reproduce the main trends of the DNPs self-assembly as previously seen experimentally [5, 8, 9, 17, 20].

4.2 Variation in liquid crystalline groups attachment

The model for DNP considered so far, is characterised by a longitudinal attachment of the terminal LC beads (**Figure 1**) and is found to self-assemble into the Sm_A and Col morphologies discussed above. With respect to the applications, the anisotropy in material mechanical properties depends on the spatial arrangement of the DNPs cores, whereas its optical properties—on the orientations of the LC beads. The Sm_A morphology is lamellar with the layers normal **a**, and also optically uniaxial, characterised by the nematic director **n** collinear to **a** (see **Figure 2**). The Col morphology is a supramolecular assembly of columns, where each column is a stack of disc-shape DNPs itself. The arrangement of columns is uniaxial and is described via vector **a**. However, there is no global nematic order of LC beads in this case, as their orientations are distributed radially in a plane perpendicular to **a**, see **Figure 4**. Therefore, optical response of the Col morphology is essentially different from that of the Sm_A one, bearing some analogy in the difference in opto-mechanical applications of the main- and side-chain LC architectures, see e.g. [21].

It is evident from **Figure 4**, that flat radial orientations of the LC beads in Col phase is the result of the energy penalty associated with bending of its host ligands. Lateral attachment of the LC bead, see, **Figure 9**, left frame, is also possible from the view of chemical synthesis [17], and may open up a possibility of greater orientation freedom of the LC beads. In particular, one may expect the optically

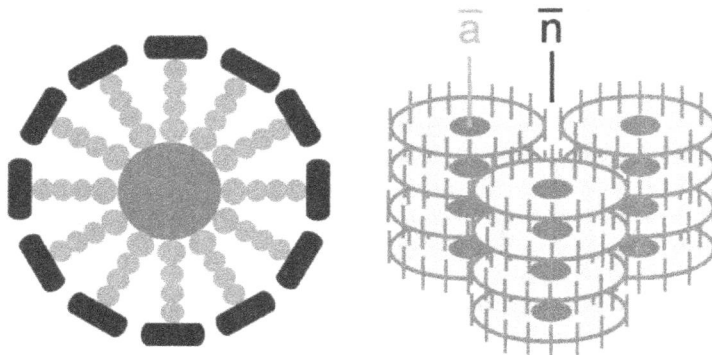

Figure 9.
Model DNP with lateral attachment of LC beads (left frame) and uniaxial Col morphology (right frame).

uniaxial columnar morphology shown in **Figure 9**, right frame, or, some other morphologies as well. With this aim we modified the DNP model accordingly. The orthogonality of LC groups to the spacer-LC bond is maintained by employing two potentials: harmonic bond between the last monomer of the spacer and the centre of the spherocylinder (the bond length is 0.3 nm) and harmonic angular potential between the latter bond and the long axis of the LC bead (equilibrium angle is $\theta_0 = \pi/2$).

Only the case of $N_{\text{ch}} = 32$ ligands is considered and the self-assembly runs are all field-assisted. The field is directed along the Z axis and its magnitude is set at $F = 2 - 4 \cdot 10^{-20}$J. The simulations are performed for the range of pressures $P = 50 - 200$ atm, the duration of each run is 10^6 molecular dynamics steps with a time step of $\Delta t = 20$ fs. For all values of the pressure within this interval the system indeed assemble into the uniaxial hexagonal columnar morphology depicted schematically in **Figure 9** (right frame) and referred thereafter as uCol$_h$. We use this morphology as an initial state and perform a series of subsequent runs at selected temperatures with the field switched off are aimed on examining the temperature stability of this phase.

For the analysis of the structural changes we consider a number of order parameters. Besides the orientation order parameter S_N (7) we also introduce the hexagonal order parameter S_H within the XY plane and columnar order parameter S_C defined as follows

$$S_H = \left\langle \left| \frac{1}{N_k} \sum_{k=1}^{N_k} e^{6j\varphi_k} \right| \right\rangle_{i,t}, \quad S_C = \left\langle \frac{N_{c,i}}{N_{\max}} \right\rangle_{i,t}. \tag{10}$$

Here φ_k is the polar angle of the bond between kth and ith DNPs, where the summation is done over all kth DNPs that belong to a first coordination sphere of ith DNP (see **Figure 10(a)**), $j = \sqrt{-1}$, $N_{c,i}$ is the number of DNPs such as their centres are found inside this cylinder of radius $R_c = 1$ nm drawn around the core of the ith DNP along the Z axis (see **Figure 10(b)**), and $N_{\max}$ is the normalisation factor introduced for the sake of convenience.

Evolutions of all order parameters, S_N, S_H and S_C upon the increase of the temperature are shown in **Figure 10**. At low temperatures, $T < 350$ K, the values of all three order parameters are essentially non-zero indicating the uniaxial hexagonal columnar phase uCol$_h$ characterised by uniaxial nematic order, high columnarity and high hexagonal order of the columns. With the increase of the temperature all

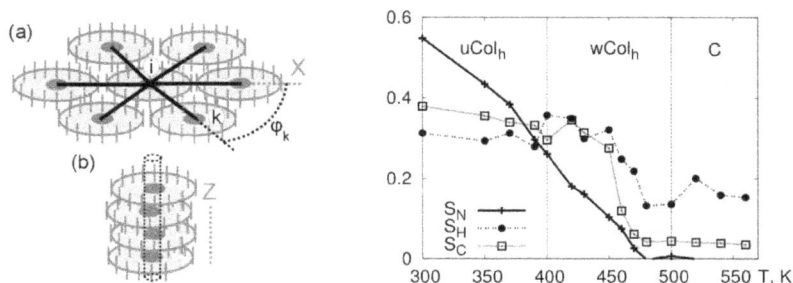

Figure 10.
(a) Flat hexagonal cluster of DNPs and the definition of bond angles φ_k. (b) A column of stacked DNPs. Order parameters S_N, S_H and S_C upon heating the uCol$_h$.

order parameters gradually decay, but while S_N decays very fast (almost linearly, in contrast with **Figure 6**), the essential delay is observed in the decrease of S_H and S_C. As the result, there is a temperature range around $T = 450$ K, where the value of S_N dropped to about 0.1 (typical for the isotropic phase), but S_H and S_C are still almost the same as in the uCol$_h$ morphology at $T = 300 - 400$ K. This is a columnar morphology with as a weak hexagonal order, wCol$_h$ but still with considerably ordered LC beads. The transition from uCol$_h$ to wCol$_h$ is gradual and the boundary between both shown in **Figure 10** is for rather illustrative purpose. With further heating of the system, all order parameters S_H and S_C drop to their minima at approximately $T \geq 480K$ indicating disordered morphology [22]. Therefore, the simulations predict two novel discotic morphologies for the DNPs with lateral attachment of LC groups: one characterised by an uniaxial, another—by random orientations of the LC groups.

5. Photo-assisted self-assembly for the azobenzene-decorated nanoparticles

Results presented in Sections 3 and 4 indicate that efficient self-assembly of model DNPs into monodomain morphologies faces certain difficulties. These are not the artefacts of the model or the simulation approach, but reflect physical properties of the LC polymers, which are relatively viscous and characterised by slow relaxation and strong tendencies for metastability attributed to the presence of a transient network [23]. One of the ways to stimulate/control the self-assembly of DNPs is to use the light [24–28]. To utilise this approach, the chromophores (e.g. azobenzene, cinnamoyl, diarylethene dithiophenols, etc.) are incorporated into the DNPs ligands. These effects share similarities with photoinduced structural changes in azobenzene-containing side-chain polymers.

Azobenzene chromophore is one of the most widely used, it exists in the trans- and cis-forms. The trans-isomer is prolate and has the LC properties, the cis-one is a non-LC. The photoisomerisation occurs between these forms when illuminated by a light with suitable wavelength. The trans-cis photoisomerization is angle-selective, the reverse one is not. If the absorption bands for both transformations overlap, the continuous trans-cis-trans isomerization cycles take place resulting in the reorientation of the trans-isomers predominantly perpendicularly to the light polarisation axis ('orientation hole-burning' or Weigert effect). Both polarised and unpolarized light can be used to photo-align the trans-isomers. In the latter case, as pointed out by Ikeda [29], 'only the propagation direction is, in principle, perpendicular to the electric vector of the light. Thus, when unpolarized light is employed, it is expected that the azobenzene moieties become aligned only in the propagation direction of the actinic light'. This situation opens up a possibility for uniaxial alignment of azobenzenes in the direction of the light propagation and is exploited for our model DNPs melt.

We follow Ref. [30] in complementing the deterministic part of the simulations, provided by molecular dynamics, with stochastic part, which describes photoisomerisation of chromophores on a coarse-grained level. We represent trans-isomers via coarse-grained t-beads and their cis-counterparts—by c-beads. Both are considered of the same spherocylinder shape, but t-t interaction is of the LC type (described by a potential (2)), whereas the t-c and c-c interactions are non-LC soft repulsive (described by a potential of the similar form as (1) and (4)). The quantum mechanical nature of photoisomerization [31, 32] is accounted for implicitly, by applying the kinetic equations of a general form for the probabilities of the transitions between the t- and c-state of each ith bead:

$$
\begin{cases}
p_i(t \rightarrow c) = p_t \left(\hat{\mathbf{e}}_i \cdot \hat{\mathbf{i}} \right)^2 \\
p_i(c \rightarrow t) = p_c,
\end{cases}
\tag{11}
$$

where $\hat{\mathbf{i}}$ is the unit vector collinear with the light polarisation axis, and p_t and p_c are the respective transition rates that depend on the chemical details of the chromophore group and the intensity and the wavelength of the illumination. The switch of the state is attempted for each chromophore at each MD step. Selective absorption of the light by the azobenzene chromophores is reflected in the angular dependence of the transition probability $p_i(t \rightarrow c)$. Photoisomerisation also involves the random change of the chromophore orientation [30, 31]. The choice being made for $p_t = 0.001$ and $p_t/p_c = 0.5$ is justified in detail in [33] and leads to the concentration of c-beads of 5–10% in a photostationary state.

As was discussed in Section 3, the model DNP is found to exhibit rod-disc shape bistability at a wide interval of $N_{\text{ch}} = 24 - 56$ (see, **Figure 8**) resulting in a polydomain mixture of both *SmA* and Col domains upon cooling it down from isotropic state. To avoid this uncertainty we use the DNPs with $N_{\text{ch}} = 12$, in which case only Sm$_A$ morphology is found.

To monitor the level of Sm$_A$ ordering in a system, we consider a set of relevant characteristics. Each DNP is considered in terms of its equivalent ellipsoid provided by the gyration tensor components (8). The asphericity, A, is defined as an average $A = \left\langle (3/2)(\lambda_{\text{max}} + \lambda_{\text{med}} + \lambda_{\text{min}})/R_g^2 - 1/2 \right\rangle$. Orientation order of DNPs is defined as $S_2 = \left\langle P_2(\hat{\mathbf{E}} \cdot \hat{\mathbf{N}}) \right\rangle$, where $\hat{\mathbf{E}}$ is the orientation of DNP long axis and the nematic director $\hat{\mathbf{N}}$ is evaluated in a way usual for LC. Global smectic order in a system is linked to the level of 'lamellarity' in the arrangement of DNPs cores and is quantified via the amplitude of the density wave along the layer normal. It is evaluated by finding the maximum of the expression

$$
S(p) = \max \left| \left\langle e^{i2\pi(\mathbf{R} \cdot \hat{\mathbf{N}})/p} \right\rangle \right|,
\tag{12}
$$

as a function of p (here $i = \sqrt{-1}$). The maximum position $p_\parallel$ provides the pitch of the SmA phase, whereas the smectic order parameter is $S_s = S\left(p_\parallel\right)$. To examine polydomain structure in a system we split it into separate Sm_A clusters (if more than one exist) and calculate their number per molecule, N_c, and the reduced maximum cluster size M_c, which is the number of DNPs in a largest cluster divided by the total number of DNPs. For the monodomain morphology one has $N_c \to 0$ and $M_c \to 1$, whereas for a highly polydomain phase: $N_c \to 1$ and $M_c \to 0$. The magnitude of all characteristics, A, S_2, S_s, N_c and M_c, are, therefore, restrained to the interval between 0 and 1.

We start from the heating runs. The initial monodomain Sm_A morphology was prepared with the aid of the orienting field (6). The changes in its properties are monitored then in a temperature interval from 400 to 550 K with the field switched off. These changes are shown in **Figure 11**. As follows from there, the system undergoes sharp changes at $T^* \approx 510\,K$: both order parameters S_2 and S_s sharply drop to zero indicating the presence of the order-disorder transition. The synchronicity in the evolution of A, S_2 and S_s indicates the absence of the purely nematic phase ($S_2 > 0$ and $S_s = 0$) and a strong relation between the molecular shape and the symmetry of the ordered morphology, similarly to the case discussed in Section 3. Therefore, the transition that occurs at T^* is the Sm_A-I transition. The behaviour of N_c and M_c indicates the monodomain morphology at $T = 400 - 500\,K$ which is transformed into a highly polydomain one in a course of a transition.

We attempt next the reverse transition: a spontaneous self-assembly of the Sm_A morphology out of the isotropic obtained upon cooling down the initial isotropic state. Various cooling rates ranging from 0.37 to 4.5 K/ns are used (note, that because of coarse-graining, the time scale is essentially contracted comparing with the real systems). The results are shown in **Figure 12**. Their behaviour demonstrate that at low enough cooling rate of 0.37 K/ns A, N_c and M_c closely follow their

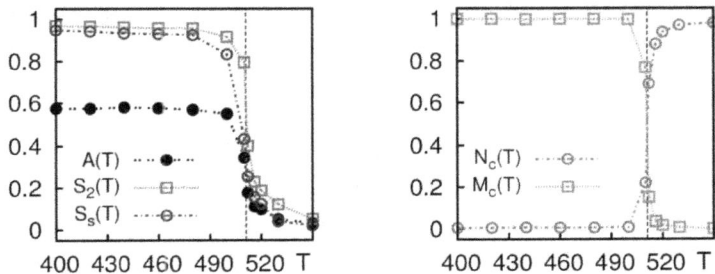

Figure 11.
A, S_2 and S_s (left frame) and N_c and M_c (right frame) vs. T. Initial state: monodomain Sm_A, heated up to T indicated in each plot. Reprinted with permission from [33]. Copyright (2016) American Chemical Society.

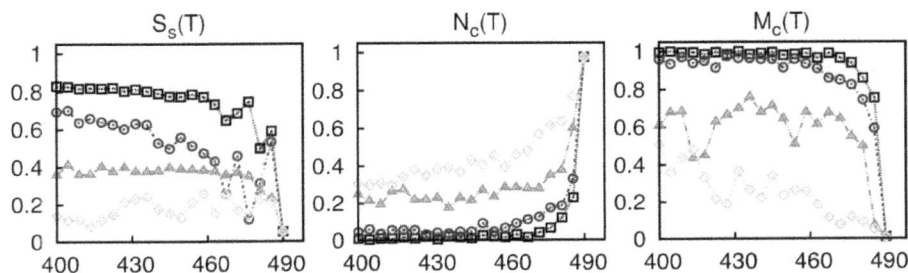

Figure 12.
Temperature dependence of S_s, N_c and M_c. Initial state: isotropic system, cooled down from 490 K to a given T. Cooling rates: Black squares: 0.37 K/ns, blue circles: 0.75 K/ns, red triangles: 1.12 K/ns, and orange diamonds: 4.5 K/ns. No illumination is applied. Adapted with permission from [33]. Copyright (2016) American Chemical Society.

respective curves shown in **Figure 11**. The order parameter S_s do not quite reach its respective value found in **Figure 11** but is, nevertheless, close. At higher cooling rate of 1.12 K/ns the system is trapped in a polydomain state with $N_c \sim 0.2$ (about 40 individual clusters). The maximum cluster size is still quite large, $M_c \sim 0.6$ (about 120 DNPs). This indicates the coexistence of one large and about 40 smaller clusters. At higher cooling rate the system is split into a larger number of smaller clusters, as indicated by N_c and M_c.

To examine the origin of this metastable state, we performed analysis of the translational and rotational mobility of DNPs in a temperature interval of 350 − 500 K. The system was quenched at each temperature for 30 ns and the initial 1 − 8 ns of each the run were analysed by splitting it into time blocks at time instances $\{t_k\}$ of equal duration $\delta t = t_k - t_{k-1} = 0.2$ ns. We define instantaneous translational and rotational diffusivities

$$d(t_k) = \frac{1}{6\delta t}\left\langle [\mathbf{R}(t_k) - \mathbf{R}(t_{k-1})]^2\right\rangle, \quad r(t_k) = \frac{1}{2\delta t}\left\langle \left[\hat{\mathbf{E}}(t_k) \cdot \hat{\mathbf{E}}(t_{k-1})\right]^2\right\rangle \tag{13}$$

at each t_k, being averaged over initial 1 − 8 ns, they provide short-time translational $D = \langle d(t_k)\rangle$ and rotational $R = \langle r(t_k)\rangle$ diffusion coefficients of the DNPs at the early stage of a self-assembly. The temperature dependence of both is shown in **Figure 13** as black legends. Both characteristics are high at 500 K and decay sharply as the temperature decreases, indicating a slowdown of translational and rotational mobility of DNPs. Hence, at lower temperatures one observes physically cross-linked domains, similarly to the case of side-chain LC polymers [23]. To quantify orientational arrest of chromophores at lower temperatures we estimated their rotation relaxation time as the function of the temperature from the time autocorrelation function $\langle \hat{\mathbf{e}}(t) \cdot \hat{\mathbf{e}}(0)\rangle$ for their orientations assuming its exponential decay

$$C(t) = \langle \hat{\mathbf{e}}(t) \cdot \hat{\mathbf{e}}(0)\rangle \sim \exp\left[-\frac{t}{t_{\mathrm{rot}}}\right], \tag{14}$$

The evaluation time interval is restricted again to the initial 1 − 8 ns of each run. The dependence of t_{rot} on temperature shows an essential increase of 1 − 1.5 order of magnitude upon lowering the temperature from 500 down to 350 K, see **Figure 13**.

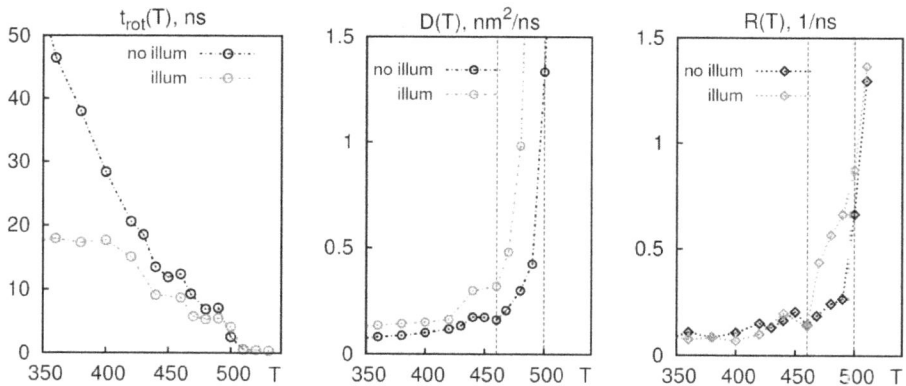

Figure 13.
Chromophores rotation relaxation time t_{rot} and short-time translational D and rotational R diffusion coefficients. Illuminated melt (red legends) vs. non-illuminated melt (black legends). Reprinted with permission from [33]. Copyright (2016) American Chemical Society.

Let us project now the effects of the illumination on the self-assembly of the DNPs. There are two known types of photomodulation in azobenzene-containing systems: (A) weakening (or elimination) of the LC order, and/or (B) order-order transition [29]. The effect (A) should weaken the interdomain links and increasing translation and rotation freedom of the DNPs. This is found in the performed simulations, as is seen from the **Figure 13** (red legends). While the differences between illuminated and no illuminated cases are negligible around 500 K, t_{rot} under illumination decreases by a factor of 2.5 at 350 K. Higher mobility of DNPs under illumination is also detected at higher valued of D and R, especially at 460 – 500 K.

These effects of illumination manifest theirselves on the self-assembly in a course of cooling runs. The results for S_s, N_c and M_c under illumination are shown in **Figure 14**. Similar behaviour is found for most characteristics as compared to the no illumination case, **Figure 12**, but at different respective cooling rates. The illumination reduces the self-assembly time-scale on average by a factor of 3–4 as compared to the no illumination case. Let us note again that the model time-scale is contracted comparing with the real units due to coarse-graining, therefore we emphasise on the relative speed-up of a self-assembly but not on the absolute cooling rates.

To have additional proof on the ability of the illumination to aid self-assembly of DNPs, we performed quenching runs. With no illumination applied, such runs end up in a glassy-like metastable state with no evidence of the orientational or positional order. The situation is markedly different under illumination, as indicated in **Figure 15** for a set of properties A, S_2, S_s, N_c and M_c. Comparing with **Figure 11** for

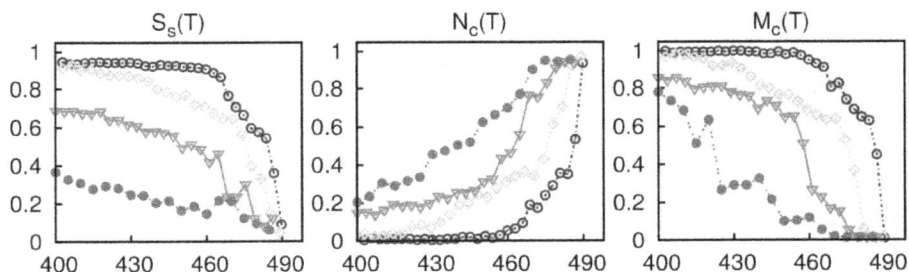

Figure 14.
Temperature dependence of S_s, N_c and M_c. Initial state: isotropic system, cooled down from 490 K to a given T indicated in the plot. Cooling rates: Blue circles: 0.75 K/ns, orange diamonds: 3.75 K/ns, pink triangles: 18 K/ns, green discs: 25 K/ns. The case with illumination. Adapted with permission from [33]. Copyright (2016) American Chemical Society.

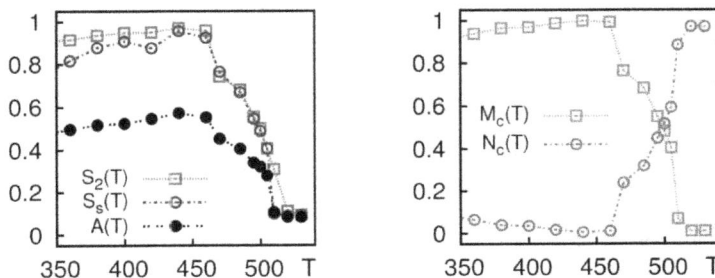

Figure 15.
A, S_2 and S_s (left frame) and N_c and M_c (right frame) vs. T. Initial state: isotropic system quenched to T under illumination. Reprinted with permission from [33]. Copyright (2016) American Chemical Society.

heating runs, one finds the curves of similar respective shapes albeit all shifted to the lower temperatures. Therefore, computer simulation studies indicate the possibility for the Sm_A phase self-assemble by quenching the system in a broad temperature interval under illumination, otherwise impossible for the non-illuminated system. Although, these simulation findings cannot be compared straightaway with the particular experiments, they are very much in-tune with the general applications of the azobenzene chromophores to control LC alignment, aggregation and self-assembly in the systems of DNPs [24–28].

6. Conclusions

Coarse-grained modelling technique maps chemical details of nanoparticles onto physical set of structure elements. In particular, the dimensions and the type of the core (metal or organic) are examined, the number and the properties of ligands (e.g. flexible or rigid, aliphatic or aromatic, etc.) are analysed, as well as the nature and properties of terminal functional groups (if any). These structure elements are replaced then by classical mechanics objects of appropriate shape that are connected via Hookean springs and interact via van der Waals forces. The parameters of the effective interaction potentials are found from related atomistic simulations using various techniques (e.g. force matching). While lacking chemical specificity of particular type of decorated nanoparticles, such models focused instead on important generic features of a whole class of underlying objects. More importantly, such simplification enables to reach required time- and length-scales of the self-assembly.

In this study we showed that such type of computer simulations is able to reproduce a wide spectra of experimentally observed features related to the self-assembly of decorated nanoparticles. Namely, we demonstrate self-assembly of model nanoparticles into lamellar smectic and columnar morphologies, reproduce and studied the shape-morphology relation, examine preference of ordered phase depending on the decoration density, study temperature driven order-disorder transitions. New optically uniaxial hexagonal phases are predicted for the case of laterally functionalised ligands.

Being viscous fluids, the melts of decorated nanoparticles are prone to slow relaxation and associated with this metastability. This restricts their application where fast regrowing/restructuring of the nanomaterial is needed in response on external stimulus. We demonstrate that the illumination of a suitable choice can be used for the speed-up of the self-assembly of the chromophore-containing nanoparticles.

The work can be extended in a number of ways. First, the use of the more accurately parameterised interaction potentials: if not for predicting the exact phase boundaries for new morphologies, then at least to show a right direction of where to search for them. Second, to consider specific (magnetic or non-metal) nanoparticles. Finally, the use of larger system sizes is always preferable. In conclusion: despite the increase in computing power, the coarse-grained picture always provides a valuable insight on the problem by focusing on its most relevant features.

Acknowledgements

The author thanks M.R. Wilson, J. Lintuvuori, M. Saphiannikova and A. Slyusarchuk for fruitful collaboration on the topic, as well as S. Sokołowski, S. Santer and E. Gorécka for stimulating discussions.

Author details

Jaroslav Ilnytskyi
Institute for Condensed Matter Physics, Lviv, Ukraine

*Address all correspondence to: iln@icmp.lviv.ua

IntechOpen

References

[1] Yokoyama M, Hosokawa K, Nogi M, Naito T, editors. Nanoparticle Technology Handbook. Amsterdam: Elsevier; 2008

[2] Hepel M, Zhong C-J, editors. Functional Nanoparticles for Bioanalysis, Nanomedicine, and Bioelectronic Devices. Vol. 1. American Chemical Society; 2012. DOI: 10.1021/bk-2012-1112

[3] Ariga K, Nishikawa M, Mori T, Takeya J, Shrestha LK, Hill JP. Self-assembly as a key player for materials nanoarchitectonics. Science and Technology of Advanced Materials. January 2019;**20**(1): 51-95. DOI: 10.1080/14686996. 2018.1553108

[4] Kumar S. Nanoparticles in discotic liquid crystals. In Series in Soft Condensed Matter. World Scientific; October 2016. pp. 461-496. DOI: 10.1142/9789814619264_0013

[5] Draper M, Saez IM, Cowling SJ, Gai P, Heinrich B, Donnio B, et al. Self-assembly and shape morphology of liquid crystalline gold metamaterials. Advanced Functional Materials. Mar 2011;**21**(7):1260-1278. DOI: 10.1002/adfm.201001606

[6] Heinz H, Pramanik C, Heinz O, Ding Y, Mishra RK, Marchon D, et al. Nanoparticle decoration with surfactants: Molecular interactions, assembly, and applications. Surface Science Reports. February 2017;**72**(1): 1-58. DOI: 10.1016/j.surfrep.2017. 02.001

[7] Agrawal AK, Kumar K, Swarnakar NK, Kushwah V, Jain S. Liquid crystalline nanoparticles: Rationally designed vehicle to improve stability and therapeutic efficacy of insulin following oral administration. Molecular Pharmaceutics. May 2017;

14(6):1874-1882. DOI: 10.1021/acs. molpharmaceut.6b01099

[8] Choudhary A, George T, Li G. Conjugation of nanomaterials and nematic liquid crystals for futuristic applications and biosensors. Biosensors. July 2018;**8**(3):69. DOI: 10.3390/bios8030069

[9] Shen Y, Dierking I. Perspectives in liquid-crystal-aided nanotechnology and nanoscience. Applied Sciences. June 2019;**9**(12):2512. DOI: 10.3390/app9122512

[10] Voth GA, editor. Coarse-Graining of Condensed Phase and Biomolecular Systems. CRC Press; 2008. ISBN 1420059556. Available at: https://www. amazon.com/Coarse-Graining-Condensed-Phase-Biomolecular-Systems/dp/1420059556?Subscription Id=0JYN1NVW651KCA56C102&tag= techkie-20&linkCode=xm2&camp= 2025&creative=165953&creative ASIN=1420059556

[11] Rühle V, Junghans C, Lukyanov A, Kremer K, Andrienko D. Versatile object-oriented toolkit for coarse-graining applications. Journal of Chemical Theory and Computation. Dec 2009;**5**(12):3211-3223. DOI: 10.1021/ct900369w

[12] Orlandi S, Zannoni C. Phase organization of mesogen-decorated spherical nanoparticles. Molecular Crystals and Liquid Crystals. May 2013; **573**(1):1-9. DOI: 10.1080/15421406. 2012.763213

[13] Baran Ł, Sokolowski S. A comparison of molecular dynamics results for two models of nanoparticles with fixed and mobile ligands in two-dimensions. Applied Surface Science. Feb 2017;**396**:1343-1351. DOI: 10.1016/j. apsusc.2016.11.159

[14] Hughes ZE, Wilson MR, Stimson LM. Coarse-grained simulation studies of a liquid crystal dendrimer: Towards computational predictions of nanoscale structure through microphase separation. Soft Matter. 2005;**1**(6):436. DOI: 10.1039/b511082c

[15] Ilnytskyi J, Lintuvuori J, Wilson MR. Simulation of bulk phases formed by polyphilic liquid crystal dendrimers. Condensed Matter Physics. 2010;**13**(3): 33001. DOI: 10.5488/cmp.13.33001

[16] Lintuvuori JS, Wilson MR. A new anisotropic soft-core model for the simulation of liquid crystal mesophases. The Journal of Chemical Physics. 2008; **128**(4):044906. DOI: 10.1063/1.2825292

[17] Saez IM, Goodby JW. Supermolecular liquid crystals. Journal of Materials Chemistry. 2005;**15**(1):26. DOI: 10.1039/b413416h

[18] Ilnytskyi JM, Neher D. Structure and internal dynamics of a side chain liquid crystalline polymer in various phases by molecular dynamics simulations: A step towards coarse graining. The Journal of Chemical Physics. 2007;**126**(17):174905. DOI: 10.1063/1.2712438

[19] Ilnytskyi J. Relation between the grafting density of liquid crystal macromolecule and the symmetry of self-assembled bulk phase: Coarse-grained molecular dynamics study. Condensed Matter Physics. 2013;**16**(4): 43004. DOI: 10.5488/cmp.16.43004

[20] Gowda A, Kumar S. Recent advances in discotic liquid crystal-assisted nanoparticles. Materials. March 2018;**11**(3):382. DOI: 10.3390/ma11030382

[21] Ikeda T, Mamiya J-i, Yu Y. Photomechanics of liquid-crystalline elastomers and other polymers. Angewandte Chemie, International Edition. Jan 2007;**46**(4):506-528

[22] Slyusarchuk A, Ilnytskyi J. Novel morphologies for laterally decorated metaparticles: Molecular dynamics simulation. Condensed Matter Physics. Dec 2014;**17**(4):44001. DOI: 10.5488/cmp.17.44001

[23] Gallani JL, Hilliou L, Martinoty P, Keller P. Abnormal viscoelastic behavior of side-chain liquid-crystal polymers. Physical Review Letters. Mar 1994; **72**(13):2109-2112. DOI: 10.1103/physrevlett.72.2109

[24] Kuang Z-Y, Fan Y-J, Tao L, Li M-L, Zhao N, Wang P, et al. Alignment control of nematic liquid crystal using gold nanoparticles grafted by the liquid crystalline polymer with azobenzene mesogens as the side chains. ACS Applied Materials & Interfaces. July 2018;**10**(32):27269-27277. DOI: 10.1021/acsami.8b07483

[25] Chu Z, Han Y, Bian T, De S, Král P, Klajn R. Supramolecular control of azobenzene switching on nanoparticles. Journal of the American Chemical Society. Dec 2018;**141**(5):1949-1960. DOI: 10.1021/jacs.8b09638

[26] Biswas TK, Sarkar SM, Yusoff MM, Rahman ML. Synthesis and characterization of azobenzene-based gold nanoparticles for photo-switching properties. Journal of Molecular Liquids. Feb 2016;**214**:231-237. DOI: 10.1016/j.molliq.2015.12.078

[27] Song H, Jing C, Ma W, Xie T, Long Y-T. Reversible photoisomerization of azobenzene molecules on a single gold nanoparticle surface. Chemical Communications. 2016;**52**(14): 2984-2987. DOI: 10.1039/c5cc10468h

[28] Wang Q, Li D, Xiao J, Guo F, Qi L. Reversible self-assembly of gold nanorods mediated by photoswitchable molecular adsorption. Nano Research. April 2019;**12**(7):1563-1569. DOI: 10.1007/s12274-019-2393-9

[29] Ikeda T. Photomodulation of liquid crystal orientations for photonic applications. Journal of Materials Chemistry. 2003;**13**(9):2037. DOI: 10.1039/b306216n

[30] Ilnytskyi JM, Saphiannikova M. Reorientation dynamics of chromophores in photosensitive polymers by means of coarse-grained modeling. ChemPhysChem. Sep 2015;**16**(15):3180-3189. DOI: 10.1002/cphc.201500500

[31] Dumont M, El Osman A. On spontaneous and photoinduced orientational mobility of dye molecules in polymers. Chemical Physics. Jul 1999;**245**(1–3):437-462. DOI: 10.1016/s0301-0104(99)00096-8

[32] Tiberio G, Muccioli L, Berardi R, Zannoni C. How does the trans-cis photoisomerization of azobenzene take place in organic solvents? ChemPhysChem. mar 2010;**11**(5):1018-1028. DOI: 10.1002/cphc.200900652

[33] Ilnytskyi JM, Slyusarchuk A, Saphiannikova M. Photocontrollable self-assembly of azobenzene-decorated nanoparticles in bulk: Computer simulation study. Macromolecules. Nov 2016;**49**(23):9272-9282. DOI: 10.1021/acs.macromol.6b01871

Chapter 3

Self-Assembly of GeMn Nanocolumns in GeMn Thin Films

Thi Giang Le

Abstract

This chapter presents the results of growing GeMn nanocolumns on Ge(001) substrates by means of molecular beam epitaxy (MBE). The samples have been prepared by co-depositing Ge and Mn at growth temperature of 130°C and Mn at concentration of ~6% to ensure the reproduction of GeMn nanocolumns. Based on the observation of changes in reflection high-energy electron diffraction (RHEED) patterns during nanocolumn growth, surface signals of GeMn nanocolumn formation have been identified. Structural analysis using transmission electron microscopy (TEM) show the self-assembled nanocolumns with core-shell structure extend through the whole thickness of the GeMn layer. Most of nanocolumns are oriented perpendicular to the interface along the growth direction. The nanocolumn size has been determined to be about 5–8 nm in diameter and a maximum height of 80 nm. A phenomenological model has been proposed to explain the driving force for self-assembly and growth mechanisms of GeMn nanocolumns. The in-plane or lateral Mn diffusion/segregation is driven by a low solubility of Mn in Ge while the driving force of Mn vertical segregation is induced by the surfactant effect along the [001] direction.

Keywords: GeMn nanocolumns, Ge thin film, growth mechanism, Mn segregation, Mn low solubility, Mn_5Ge_3 clusters

1. Introduction

The discovery of the giant magneto-resistance (GMR) effect in metallic multilayers by Albert Fert and Peter Grunberg has probably made a great step toward spintronics [1, 2]. To further extend applications of spintronics, researchers and engineers invent new architectures and structures, allowing to realize the integration of magnetic materials into semiconductors. The development of active spin devices, such as spin transistors or diodes, calls for new materials, which enable to efficiently inject spin-polarized currents into standard semiconductors. Two main ways have been explored in order to inject spin-polarized current into semiconductors.

Firstly, one can make use of the properties of a ferromagnetic metal (FM) such as Co, Fe, Ni, or their alloys. Spin-polarized currents tunnel from ferromagnetic metal to semiconductor through an insulator [3, 4] or a Schottky barrier [5]. However, the efficiency of spin injection directly into Si or Ge remains very low. Indeed, most of the ferromagnetic metals react with Si and Ge, leading to the formation of interfacial silicides or germanides, which, for most of them, are not ferromagnetic. It is also not trivial to obtain epitaxial growth of an oxide in between Ge (or Si) and a ferromagnetic metal; spin injection is therefore limited by the interface roughness [6].

Secondly, diluted magnetic semiconductors (DMSs), obtained by doping standard semiconductors with magnetic impurities, such as Mn or Co, have emerged as potential candidates for spin injection. The materials become ferromagnetic while conserving their semiconducting properties. They exhibit therefore natural impedance match to host semiconductors and are expected to efficiently inject spin-polarized currents into semiconductors. Since the spintronic devices often operate at room temperature and they are heated up during the operation, the great challenge and ultimate goal of the research in this field is to obtain DMSs exhibiting ferromagnetism well above room temperature. This feature represents key issues for the development of spintronic devices.

In the 1990s, DMSs of III-V-based compound semiconductors were successfully fabricated by introducing Mn ions, but Mn atoms are much less soluble than in II-VI semiconductors, making them difficult to be diluted in the III-V semiconductor, such as (GaMn)As. By using a low-temperature MBE technique, it is possible to grow thin films with higher Mn concentrations in a nonequilibrium process and prevent Mn ions to form precipitations. The T_C, achieved at that time, was 110 K for 5.5% Mn-doped GaAs [7]. So far, the GaMnAs diluted magnetic semiconductors seem to be the most important and the best understood system up to now. However, they are ferromagnetic only at temperatures well below room temperature, the highest value reported was 173 K by Gallagher's group in UK [8]. An interesting alternative could be magnetic semiconductors that are based on elemental semiconductors and also owe to their compatibility with Si microelectronics. In the last decades, considerable amount of work has been devoted to the synthesis of Mn-doped Ge and Si, such as SiMn, GeMn, and SiGeMn. The main motivations for the synthesis of these materials are:

- Compatibility with mainstream silicon technology.

- Mn magnetic impurity acts as an acceptor in the substitutional sites in the crystal lattice.

- Very long spin relaxation time, which comes from weak spin-orbital coupling in Si and Ge [9]. It is worth noting that the spin relaxation time in IV-IV semiconductors is much larger than that in III-V semiconductors.

Although silicon is the key material of microelectronics, the first demonstration of spin injection was only achieved in 2007. Until now, it is unclear whether Mn can substitute Si sites since Mn ions in Si are fast interstitial diffusers even at low temperatures. Experimentally, numerous groups have reported the observation of ferromagnetism in Mn-doped Si with Curie temperatures ranging from 200 to 400 K [10–13]. However, the origin of the observed ferromagnetism remains very diverse, which makes Mn-doped Si DMSs difficult to be realized.

Recently, special attention both in experiment and theory has been given to group-IV $Ge_{1-x}Mn_x$ diluted magnetic semiconductors due to their compatibility with mainstream Si-based electronics. Zwicker et al. conducted the first study on the Ge-Mn equilibrium phase in the early 1949 [14]. Later on followed extensive investigations establishing the relative phase diagram [15]. The interests in GeMn system as a promising high-T_C ferromagnetic semiconductor essentially started in 2002 with the publication by Park et al. [16], claiming the fabrication of a GeMn DMS with a T_C up to 116 K and linear dependence on the Mn concentration. Since then, many publications investigated different aspects of the GeMn system and employed different fabrication techniques [17–19].

Despite numerous researches carried out up to now, the fabrication of homogeneous and high-T_C $Ge_{1-x}Mn_x$ films remains a challenge. It is now commonly believed that secondary phases form once the Mn concentration exceeds the solubility limit [20]. In general, when the Mn concentration and the growth temperature are high enough or after post-annealing, parasitical metallic nanoclusters of Mn_5Ge_3 are observed in GeMn films [21–23]. $Mn_{11}Ge_8$ precipitates are observed under certain conditions [24]. A result of particular interest is the discovery of the CEA group on the formation of a nanocolumn phase, which exhibits a Curie temperature higher than 400 K [25]. The nanocolumns have an average size of 5 nm and are separated from each other by a distance of about 8 nm. This phase has been attributed to a new compound, Ge_2Mn, which does not exist in the Ge-Mn phase diagram.

Among numerous phases of GeMn DMS, the nanocolumns' phase appears to be the most interesting, because it is the unique phase that has Tc higher than RT. But concerning the GeMn nanocolumns, there are some remaining questions about the composition and the mechanism of nanocolumn formation. One of the main objectives of this chapter consists of understanding the origin of the formation of this nanocolumn phase. Our purposes are to determine the nanocolumn size, driving force for self-assembly, and growth mechanisms of GeMn nanocolumns.

2. Literature review of the synthesis of $Ge_{1-x}Mn_x$ DMS

After the first report of ferromagnetism in the $Ge_{1-x}Mn_x$ system with $T_C = {\sim}116$ K [16], the synthesis of $Ge_{1-x}Mn_x$ DMSs has been the subject of numerous investigations [26–36]. Since it has been shown in Ref. 10 that magnetic ordering in $Ge_{1-x}Mn_x$ linearly increases with increasing Mn concentration, a number of works have focused on investigations of the dependence of the Curie temperature of GeMn DMS on the growth parameters. Among numerous growth parameters, which can affect the $Ge_{1-x}Mn_x$ growth behavior, the growth temperature and the Mn concentration appear to be the most important. Before presenting our results, we summarize below some major results, reporting on the influence of these two parameters on the growth and magnetic properties of GeMn DMSs.

2.1 Effects of Mn concentration on the growth behavior of $Ge_{1-x}Mn_x$ DMS

Similar to the case of Mn in III-V semiconductors, the solid solubility of Mn in Ge is as low as 10^{15} cm^{-3} under equilibrium conditions [37]. As a consequence of such a low solubility, thin film growers use low-temperature growth to bring the system to high nonequilibrium conditions with the hope to increase the Mn concentration that can be incorporated in the films. Low-temperature growth is also expected to minimize phase separation and/or the formation of unwanted compounds. However, it is worth noting that low-temperature growth often conducts to the formation of metastable state and low crystalline quality of epilayers due to a low surface diffusion coefficient of adatoms on surface. Moreover, the ratio between interstitial and substitutional dopants may be reduced due to a too low growth temperature.

Several reports have shown that a concentration of holes up to 10^{17}–10^{20} cm^{-3} can be obtained when doping Mn in Ge [16, 25, 29, 30, 38, 39]. It has also been demonstrated that when Mn atoms substitute Ge sites, Mn ions are in 2+ ionic states [35, 36, 40, 41]. Li et al. [38] reported on Mn ionic implantation in Ge and suggested that up to 20–30% Mn ions could be incorporated in substitutional sites and post-annealing allowed increasing this fraction to 40–50%. Some interstitial Mn ions can be converted to substitutional Mn under adequate post-annealing.

In general, it is now generally accepted by the scientific community working on GeMn DMSs that a Mn concentration up to 5–9% can be incorporated in Ge layers without creating visible metallic clusters or precipitates, such as Mn_5Ge_3 and $Mn_{11}Ge_8$. However, other kinds of Mn-rich cluster phases, in particular with size being in the range of sub-nanometer, may exist [41–43]. A question, which is still under debate, concerns the exact Mn amount or concentration that really participates in producing ferromagnetic ordering in $Ge_{1-x}Mn_x$ materials. Answering this question certainly requires nano-scaled characterization tools for both structural and magnetic properties. Another remaining question concerns the origin of ferromagnetism in $Ge_{1-x}Mn_x$. If the Zener model or hole-mediated ferromagnetic model, which uses thermodynamic and micro-magnetic descriptions to explain the origin of ferromagnetism in DMS materials, is generally accepted, other mechanisms, for example bound magnetic polarons, have been proposed [36]. These authors, by combining magnetic and transport characterizations of $Ge_{1-x}Mn_x$ films doped with ~5% of Mn, provided the existence of two magnetic transitions: T_C^* (~112 K) and T_C (~12 K) of the layers. The transition at high temperature (T_C^*) can be attributed to the exchange interaction between localized charge carriers and surrounding Mn ions (magnetic polarons) while the T_C transition may arise from infinite-size ferromagnetic clusters due to inhomogeneous distribution of Mn dopants inside the layers.

2.2 Effects of growth temperature on $Ge_{1-x}Mn_x$ DMS

Together with the Mn concentration, the growth temperature is recognized as one of the main parameters that governs the growth process of GeMn films. In particular, the growth temperature has a direct consequence on the nature of phase separation or clusters that are formed in the layers: nano-scaled Mn-rich regions or metallic clusters (Mn_5Ge_3, $Mn_{11}Ge_8$ or Mn_5Ge_2). According to previous works, three main regions of the growth temperature can be classified: (i) above 180°C, (ii) below 80°C, and (iii) intermediate temperatures in the range from 110 to 150°C.

The most "famous" or the most frequently observed Mn-Ge clusters are certainly metallic Mn_5Ge_3 clusters. This phase of clusters is generally observed when the growth temperature exceeds 180°C. **Figure 1** shows an example in which a high density of Mn_5Ge_3 clusters is present in $Ge_{1-x}Mn_x$ films, which were grown at 225°C and contained a Mn concentration of 3% [23]. Using transmission electron diffraction (TED) and magnetic characterizations, the authors have unambiguously identified that those clusters are made of the Mn_5Ge_3 phase. An epitaxial relationship between these clusters and Ge(001) has been determined: a majority of them are oriented with the hexagonal (0001) plane of Mn_5Ge_3 being aligned with the (001) plane of Ge while some others are randomly distributed in the layers (see **Figure 1**) (right).

In the low growth temperature range, Bougeard et al. [42] reported that $Ge_{1-x}Mn_x$ films were free of intermetallic precipitates but a new kind of cluster was formed. **Figure 2a** shows a plan-view TEM image of a $Ge_{1-x}Mn_x$ film grown at 60°C with Mn content of 5%. Numerous areas exhibiting slightly darker contrast are observed; these areas were found to be coherently bounded to the surrounding Ge matrix. Analyses performed with annular dark field by scanning transmission electron microscopy (STEM) on a single cluster string (**Figure 2b**) revealed high structural disorder around the clusters, suggesting that they have a strained shell

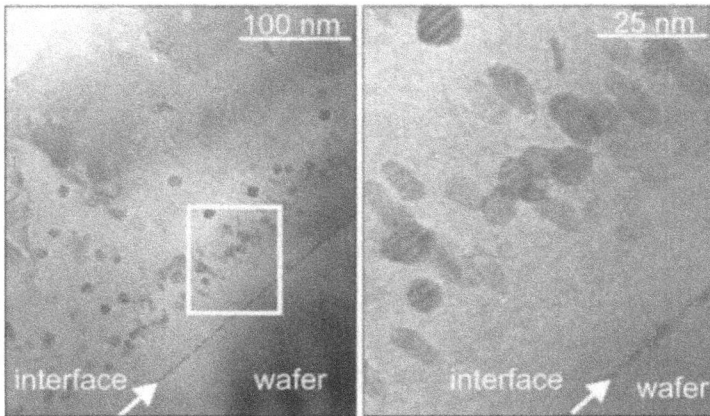

Figure 1.
TEM images of a $Ge_{0.97}Mn_{0.03}$ epilayer on a Ge wafer (left) with a magnified section (right). The arrows mark the interface between the wafer and epilayer [27].

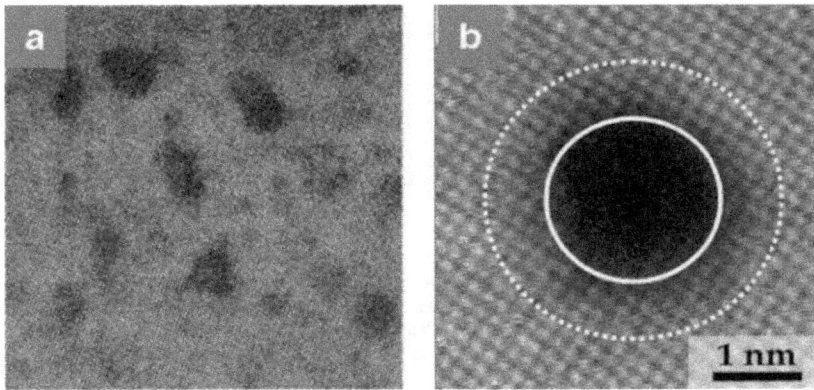

Figure 2.
High-resolution plan-view STEM images of $Ge_{0.927}Mn_{0.073}$ sample grown at 60°C. Mn-rich regions with darker contrast are amorphous Mn [44].

with a core made up by amorphous Mn [32]. Continuing their investigations, the authors recently showed that when the growth temperature increased to 85–120°C, metallic Mn_5Ge_3 in the shape of nanometer-sized inclusions can be formed in a diluted $Ge_{1-x}Mn_x$ matrix [45]. It is worth noting that when post-annealing was carried out on $Ge_{1-x}Mn_x$ films grown at the temperature range of 70–120°C, new kinds of clusters, $Mn_{11}Ge_8$ and Mn_5Ge_2 can be formed near the surface region.

A result of particular interest, observed in the intermediate temperature range from 110 to 150°C, is the observation by Jamet and colleagues at CEA-Grenoble of highly elongated precipitates, which are self-organized to form nanocolumns [18]. As can be seen in the cross-sectional TEM image shown in **Figure 3a**, nanocolumns that are elongated along the whole GeMn layer are observed. A high-resolution TEM image shown in **Figure 3b** indicates that nanocolumns are coherent with the surrounding matrix and they have an average size of about 3–5 nm. By using electron energy-loss spectroscopy (EELS), the authors determined an average Mn concentration in nanocolumns ranging from 32 to 37.5% and attributed to a Ge_2Mn alloy, which does not exist in the Ge-Mn phase diagram.

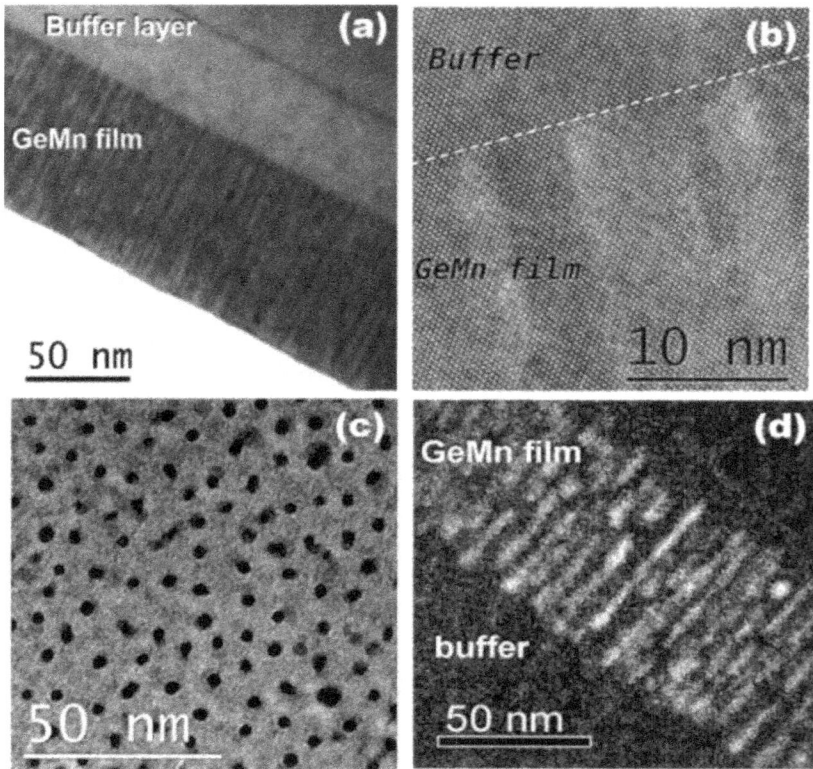

Figure 3.
Transmission electron micrographs of a Ge1–xMn$_x$ film grown at 130°C and containing 6% of manganese. (a) Cross section along the [110] axis; (b) High-resolution image of the interface between the Ge$_{1-x}$Mn$_x$ film and the Ge buffer layer; (c) Plane view micrograph performed on the same sample; (d) Mn chemical map obtained by energy-filtered transmission electron microscopy. Bright areas correspond to Mn-rich regions [39].

One of the most interesting features of this nanocolumn phase is that it exhibits magnetic ordering well above 400 K. The temperature dependence of magnetization of the corresponding sample is shown in **Figure 3a** of Ref. [25]. Magnetization of the layer persists at a temperature of 400 K, the upper limit of the instrument for magnetic measurements. Shown in the insert is the M(T) measurement after subtracting the magnetic signal of nanocolumns, much lower magnetic ordering is observed and has been attributed to a Mn-poor matrix between nanocolumns. It is worth noting that the above nanocolumns were shown to be stable up to 400°C and transformed into Mn$_5$Ge$_3$ clusters after 15 min of annealing at 650°C. This feature represents as a starting point of our research; our purpose consists of seeking the ways to stabilize this high-T$_C$ nanocolumn phase.

3. Experiment detail

The solubility limit of Mn in Ge is very low (estimated to be 10^{15} cm^{-3} [37], corresponding to a Mn atomic concentration of $\approx 2\times10^{-6}$ %). While using nonequilibrium growth techniques such as low-temperature MBE growth, a much higher Mn solubility has been demonstrated. However, such a low thermodynamical Mn solubility constitutes as the main obstacle to get homogeneous highly doped GeMn films. On the other hand, the Mn dilution in a germanium crystal remains highly questionable and in most of experiments reported up to now; GeMn films are in general found to contain either nanoscaled Mn-rich regions or secondary-phase precipitates like Mn_5Ge_3. This means that it is crucial to combine nanoscale characterizations of the structural properties with magnetic properties in order get an accurate physical picture of grown films. The ultimate goal of research in GeMn materials could be thus find out growth techniques and growth conditions to obtain homogenous materials containing a few percent of Mn, allowing to raise the Curie temperature well above room temperature for the realization of spintronic devices. Up to now, low-temperature MBE technique has been intensively used to synthesize GeMn DMSs and promising results have been obtained [32, 33, 44, 46–48].

In our works, GeMn film growth was performed using solid source molecular beam epitaxy (MBE) on epi-ready Ge(001) wafers with a base pressure better than 1×10^{-10} Torr. Ge and Mn were evaporated using standard Knudsen effusion cells. The deposition chamber is equipped with RHEED to control the sample surface and to monitor the epitaxial growth process. The Ge deposition rate was estimated from RHEED intensity oscillations of the Ge-on-Ge homoepitaxy, whereas the Mn concentrations were deduced from Rutherford backscattering spectrometry (RBS) measurements. The cleaning of the Ge surfaces was carried out in two steps: the first was a chemical cleaning to eliminate hydrocarbon contaminants. Then, the Ge native oxide layers were etched in a diluted hydrofluoric acid solution (10%) for some minutes until a hydrophobic surface was obtained. The second step was an *in situ* thermal desorption of the surface oxide, which consists of outgassing the sample for several hours at 450°C followed by flash annealing at 750°C. After this step, a 40-nm-thick Ge buffer layer was grown at 250°C to ensure a good starting surface. Regarding the previous results, 80-nm-thick $Ge_{1-x}Mn_x$ films were subsequently grown at the substrate temperature of 130°C and Mn concentrations of ~6% to ensure the reproduction of GeMn nanocolumns.

Structural analyses of the grown films were performed through extensive high-resolution transmission electron microscopy (HR-TEM) by using a JEOL 3010 microscope operating at 300 kV with a spatial resolution of 1.7 Å. Imaging at this resolution requires TEM samples to be very thin in the region of interest. As a minimum requirement, samples must be electron transparent. To take advantage of a TEM's high resolution, it is necessary to prepare high-quality electron transparent samples. In practice, TEM specimens should be no thicker than 100 nm for low-resolution imaging and even thinner (~50 nm) for high-resolution imaging [49]. Sample preparation was carried out by standard mechanical polishing and argon ion milling. At first, the samples were cut in in two rectangular pieces (about 2 × 3 mm) and the film sides glued together with epoxy. Then, the samples were polished to a thickness of approximately 20 μm with a series of progressively smoother diamond papers. Gatan PIPS™ (Precision Ion Polishing System) ion mill was used for final thinning to electron transparency.

4. The surface signal of GeMn nanocolumn formation

The surface signal of nanocolumn phase is extremely important because this allows us to know if the nucleation of the nanocolumn phase takes place. **Figure 4** displays the evolution of RHEED patterns taken along the [1-10] azimuth of Ge surface during the growth of $Ge_{0.94}Mn_{0.06}$ film. Starting from RHEED pattern of the Ge surface prior to growth in **Figure 4a**, the Ge surface is characterized by a well-developed Ge reconstruction (2 × 1) streaky pattern, indicating that the surface is clean and smooth. This is also confirmed by the observation of high-intensity

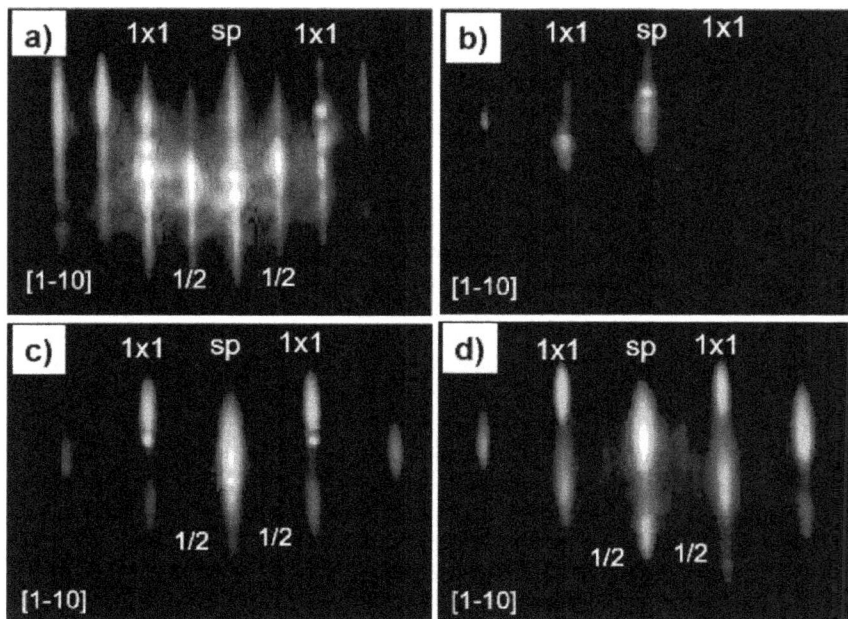

Figure 4.
RHEED patterns taken along the [1-10] azimuth of the clean Ge surface prior to growth (a), during the growth of the first monolayers (b), after 3 minutes of the growth (c), and finishing the growth process of ~80 nm $Ge_1 - _xMn_x$ film (d), with Mn concentration ~ 6%.

Kikuchi lines, which overlap 1 × 1 and specular streaks, giving rise to localized reinforcements of intensity. The calculations of density functional theory (DFT) are in good agreement with these results and demonstrate that in the case of the clean surface Ge (001), the total energies are minimal for this type of symmetry [50]. The Ge reconstruction streak (2 × 1) disappears after a few monolayers of GeMn co-deposition (**Figure 4b**), then reappears in **Figure 4c**. The high-intensity spots due to the diffraction in transmission begin to appear on the 1 × 1 streaks after the deposition of a few nanometers and remain present until the end of the deposition (**Figure 4c** and **d**). This indicates the formation of a new GeMn phase having the same crystal structure as that of the Ge substrate. In addition, the surface of this new phase is rough. These three-dimensional (3D) spots come from the contribution of the nanocolumns that are formed in a diluted GeMn matrix whose structure is similar to that of Ge. Indeed, during homoepitaxy of Ge on Ge at this same growth temperature, the RHEED pattern presents diffraction streaks characteristic of two-dimensional growth. The observation this type of pattern is very interesting because it indicates that the growth surface is rough and that the degree of roughness remains constant during the deposition. In other words, the vertical growth speed of the area between the nanocolumns and that of the nanocolumns themselves are almost the same because the diffraction diagram remains unchanged. In addition, by comparing with the TEM images obtained, it is pointed out that the diameter of nanocolumns is smaller than the coherence length of the incident electron beam (10 nm), which makes it possible to attribute unambiguously the 3D spots on the 1 × 1 streaks of surface diffraction to the presence of the nanocolumn phase.

In short, in case of nanocolumn formation, the surface still exhibits a two-dimensional growth behavior although some intensity reinforcements at the Bragg diffraction positions are visible and the intensity of reconstructed ½ streaks becomes weaker—the sign of GeMn nanocolumn formation of [51]. The 3D spots situated on the 1 × 1 streaks are the contribution of Mn-rich nanocolumns formed in the diluted GeMn matrix.

5. The growth direction and size of the GeMn nanocolumns

Figure 5 displays typical cross-sectional TEM images of a sample grown at 130°C and with Mn content of ~6%. Dark contrast corresponds to Mn-rich regions while regions with a brighter contrast arise from the diluted matrix. The corresponding film thickness is ~80 nm. According to an overall view of the layer structure, shown in low-scaled images in **Figure 5(a)**, we can see that the GeMn nanocolumns observed are very similar to those reported in Ref. 19. The self-assembled nanocolumns extend through the whole thickness of the GeMn layer. Most of nanocolumns are oriented perpendicular to the interface, that is, along the [001] direction. High-resolution TEM image taken around the nanocolumn inside the layer in **Figure 5(b)** reveals that nanocolumns are epitaxial and perfectly coherent with the surrounding diluted lattice. No defects nor presence of MnGe clusters are visible. **Figure 5(c)** shows that the interface is almost invisible and GeMn film exhibits the same diamond structure as Ge buffer layer. According to our previous study, at the growth temperature of Mn concentration of 6%, the nanocolumn can reach the maximum length of 80 nm before transforming into Mn_5Ge_3 clusters [52–54].

To investigate the arrangement and the size of GeMn nanocolumns, **Figure 6** displays the overall and high-resolution plan-view TEM images of 80-nm-thick $Ge_{0.94}Mn_{0.06}$ film. The nanocolumns are distributed evenly on the entire film surface with the diameter of about 5–8 nm. The inter-distance between adjacent columns is about 8–10 nm. The circle shape of Mn-rich nanocolumns in plan-view TEM images means that the GeMn nanocolumns exhibit cylindrical shape. Around each column in **Figure 6b**, a dark ring reveals a large strain extending over a few interatomic distances. The presence of a disordered core in the columns can reveal a plastic relaxation of the misfit stress with the surrounding matrix, probably due to the high-energy electron beam of the microscope. Studying the influence of Mn content

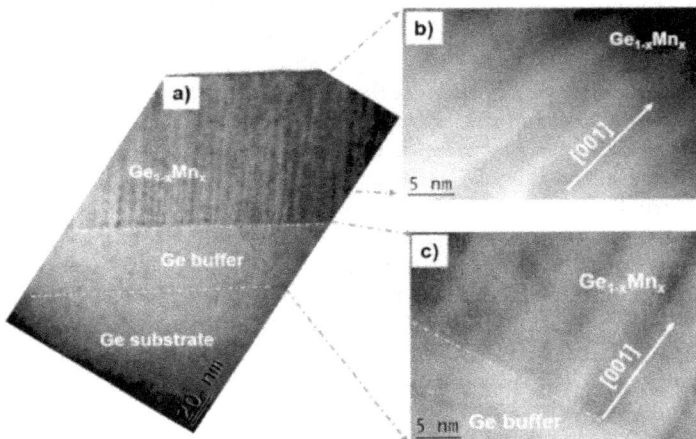

Figure 5.
Typical cross-sectional TEM image of a 80-n-thick $Ge_1 − {}_xMn_x$ film with x ~ 0.06 (a), high-resolution TEM image taken inside the film (b) and around the interface region (c).

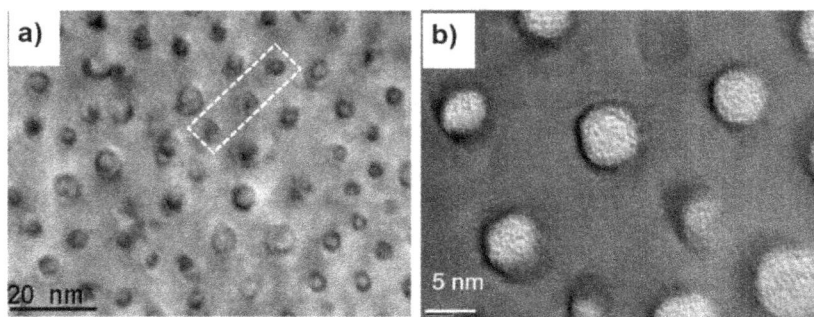

Figure 6.
Typical (a) and high-resolution (b) plan-view TEM image of a 80-nm-thick $Ge_{1-x}Mn_x$ film with x ~ 0.06

on the formation of GeMn nanocolumns, previous studies show that increasing the Mn content leads to the column density remaining nearly constant, whereas their width increases and their height decreases drastically [39, 53].

In summary, GeMn nanocolumns are found to be formed at the Mn concentration of ~6%. We are thus able to stabilize the nanocolumn phase, which extends through the whole 80-nm thickness of the GeMn layer and exhibits the diamond structure elongated along the growth direction, with the diameter of about 5–8 nm.

6. Driving force for self-assembly and growth mechanisms of GeMn nanocolumns

The incorporation of impurities in a crystal lattice generally induces a constraint either at the local scale or in the whole lattice. With a very small amount of Mn (~6%), large-scale deformation is not expected. Since the atomic radius of an Mn atom (140 pm) is larger than that of a Ge atom (125 pm), the Mn-rich region, such as nanocolumns, should be in compression because of the presence of the Ge matrix surrounding them. The RHEED technique is very local but is not sufficient to be able to observe an infinitesimal constraint induced by nanocolumns. A Fourier transform of the cross-section TEM image inside the GeMn nanocolumn layers displayed in **Figure 7** shows that no dislocation can be observed in the filtered image along the Bragg's spots (220) of the red square area. This result indicates that the nanocolumn is in a stressed state. Note that filtering by the Fourier transform of the entire nanocolumn does not reveal any dislocation. It means that the nanocolumns are coherently strained in compression by the matrix along the growth direction. In any case, the presence of stress is inevitable even if the quantity of Mn incorporated is extremely small. The stress can be the important effect in the distribution of Mn atoms in the Ge matrix. A detailed study of the constraints would make it possible to better understand the mechanism of incorporation of Mn into nanocolumns and therefore to better understand the formation and growth kinetics of nanocolumns.

One of the central results concerning the nanocolumn formation is that the Mn concentration inside nanocolumns is not constant but increases from the interface to the film surface [55]. To understand the variation of the Mn concentration inside the $Ge_{1-x}Mn_x$ nanocolumns, we attempt to provide a phenomenological explanation of the $Ge_{1-x}Mn_x$ nanocolumn formation.

To investigate the mechanism leading to the formation of $Ge_{1-x}Mn_x$ nanocolumn, we recall that under nonequilibrium growth carried out at a temperature as low as 130°C, the Ge lattice can dilute a Mn concentration of about 0.25–0.5% [56].

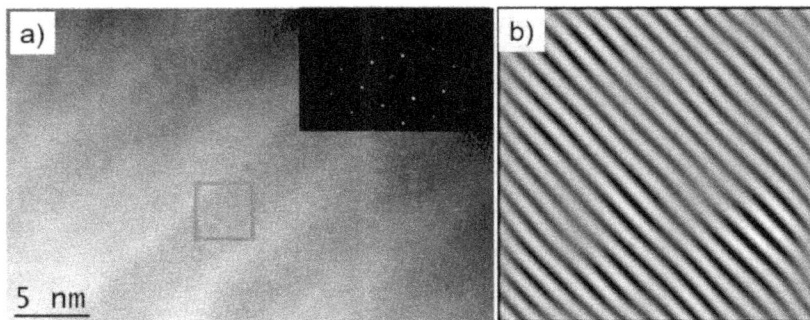

Figure 7.
High-resolution cross-section TEM image of the nanocolumns (a) and the image filtered along the Bragg's peak (220) of the red square area (b).

At the first stage of the growth, GeMn alloys with Mn concentration as high as ~6% are deposited on a Ge surface. This content far exceeds the solubility threshold of Mn atom in the Ge lattice. The excess Mn should diffuse and/or segregate along the surface to form Mn-rich regions. In other words, Mn-rich regions start to nucleate on the Ge surface even after deposition of the first monolayers and this is a direct consequence of the low solubility of Mn in the Ge lattice. The formation of these Mn-rich regions on the surface should "disturb" the surface morphology. This is what we have observed from RHEED patterns in **Figure 4b**. Probably, the presence of Mn-rich nuclei is responsible for the change in RHEED pattern of **Figure 4a** to RHEED pattern of **Figure 4b**. Since a too high Mn concentration in a nucleus is not favorable, both from the thermodynamical and epitaxial points of view, the system should self-organize to form nuclei with a certain density. The size of nuclei and the distance between nuclei depend on the Mn diffusion length on the film surface and the substrate temperature. Schema in **Figure 8** illustrates the formation of these Mn-rich nuclei on the Ge surface. During the GeMn deposition process, these nuclei may serve as seeds for the nanocolumn formation. **Figure 8** displays a schema illustrating the Mn accumulation on the Ge(001) surface.

The increase in Mn concentration within nanocolumns could be explained by vertical segregation of Mn along the [001] direction. The energy calculations along the Ge(001) orientation using spin-polarized density functional theory (DFT) show that

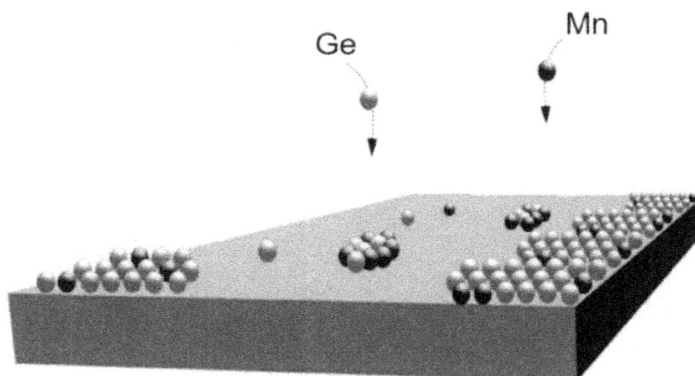

Figure 8.
The scheme of Mn accumulation on Ge(001) surface.

for Ge(100) -2×1 surface reconstruction, Mn diffuses via the surface interstitial site (I_0 site) preferentially. Mn adatoms originating from the gas phase or from deeper layers can easily diffuse toward the interstitial sites right beneath the dimers of Ge(100) -2×1 surface reconstruction. Therefore, growth of Mn-doped Ge(100) DMS at low temperatures should result in a high density of interstitial Mn. As Mn atoms are buried beneath a newly deposited Ge layer, they tend to float upward via the I_0 sites [44, 57]. Due to this surfactant effect of Mn atoms along the [001] direction, once Mn-rich nuclei are formed on the surface, further deposition leads to the formation of GeMn columns in which the Mn concentration continuously increases from interface to the film surface. And the cylindrical shape of nanocolumns allows them to minimize their interface energy with the surrounding diluted matrix. **Figure 9** displays a schema illustrating the segregation of Mn atoms during the growth GeMn nanocolumns.

According to our previous studies, the formation of GeMn nanocolumns and Mn_5Ge_3 clusters is a competing process [52]. During the growth, Mn continuously segregates toward the film surface and GeMn nanocolumns are found to transform to metallic Mn_5Ge_3 precipitates once the Mn concentration inside nanocolumns exceeds a highest value about 40%. It means that depending on the Mn concentration in the top of nanocolumns, nanocolumn growth can be interrupted. In the same time, other nanocolumns can start to nucleate in the middle of the layer if these regions are rich

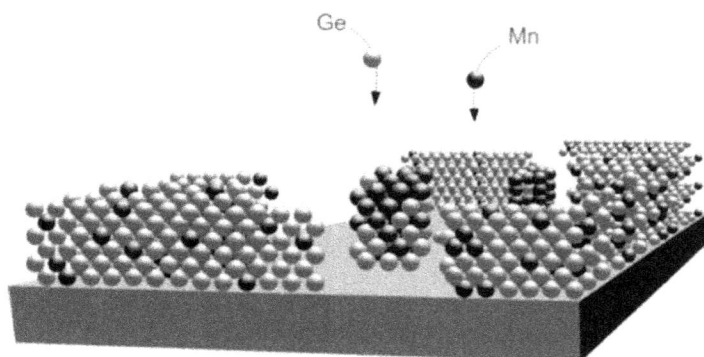

Figure 9.
The scheme of Mn atoms segregating toward the film surface through the interstitial sites during the deposition of $Ge_{1-x}Mn_x$ film on Ge(001) substrate.

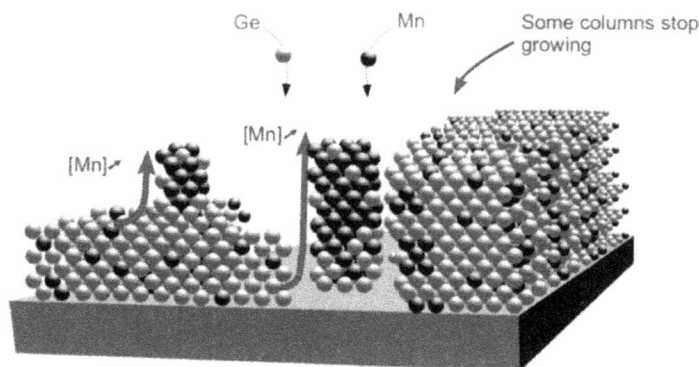

Figure 10.
The scheme of growth competition between $Ge_{1-x}Mn_x$ nanocolumns and Mn_5Ge_3.

enough in Mn to form new nuclei. **Figure 10** displays a schema illustrating the growth competition between $Ge_{1-x}Mn_x$ nanocolumns and Mn_5Ge_3 clusters.

Another interesting feature of nanocolumns is their core-shell structure [52, 53]. The Mn concentration across a nanocolumn is not homogenous; nanocolumns exhibit a core-shell structure with a much higher Mn concentration in the core compared to that of the shell. The Mn concentration in GeMn nanocolumns is highly inhomogeneous, it increases from interface to the film surface and also from the shell to the core of nanocolumns. Since the atomic radius of Mn atoms is slightly larger than that of Ge (127 and 122 picometers for Mn and Ge atoms, respectively), if the Mn concentration inside nanocolumns is too high, nanocolumns may exert a tensile strain on the surrounding lattice. To reduce such a strain, a core-shell structure with a reduced Mn concentration from core to shell should allow nanocolumns to more easily adapt the lattice parameter of the surrounding lattice. Thanks to this self-organized core-shell structure, almost all nanocolumns are found to be perfectly coherent with the surrounding lattice, as can be seen in **Figure 5**.

7. Conclusion

In conclusion, GeMn nanocolumns elongated along the growth direction are found to be formed at the growth temperature of 130°C and Mn concentration of ~6%. Based on the surface signals of GeMn nanocolumn formation, we are thus able to stabilize the nanocolumn phase, which exhibits the same diamond structure as Ge substrate with the diameter of about 5–8 nm and the maximum height of 80 nm. We have attempted to explain the nanocolumn formation using a phenomenological model based on Mn segregation and diffusion both in-plane and along the growth direction: (i) Mn-rich regions start to nucleate on the Ge surface even after deposition of the first monolayers and this is a direct consequence of the low solubility of Mn in the Ge lattice; (ii) due to the surfactant effect of Mn atoms along the [001] direction, further deposition leads to the formation of GeMn nanocolumns in which the Mn concentration continuously increases from interface to the film surface; (iii) GeMn nanocolumns become unstable when the Mn concentration reaches a value of 40% and then transform into Mn_5Ge_3 clusters. The shell-core structure of the column is formed to reduce strain, which is induced due to the difference in atomic radius between Ge and Mn.

Acknowledgements

The author thanks colleagues at Interdisciplinary Center of Nanoscience of Marseille (CINaM-CNRS), Aix-Marseille University, France for their technical help and fruitful discussions.

Author details

Thi Giang Le
Hong Duc University, Thanh Hoa, Vietnam

*Address all correspondence to: giangle74@gmail.com

IntechOpen

References

[1] Wang R, Jiang X, Shelby RM, Macfarlane RM, Parkin SSP, Bank SR, et al. Label-free detection and classification of DNA by surface vibration spectroscopy in conjugation with electrophoresis. Applied Physics Letters. 2005;**86**:052901. DOI: 10.1063/1.1853529

[2] Jiang X, Wang R, Shelby RM, Macfarlane RM, Bank SR, Harris JS, et al. Highly spin-polarized room-temperature tunnel injector for semiconductor spintronics using MgO(100). Physical Review Letters. 2005;**94**:056601. DOI: 10.1103/PhysRevLett.94.056601

[3] van't Erve OMJ, Kioseoglou G, Hanbicki AT, Li CH, Jonker BT, Mallory R, et al. Comparison of Fe/Schottky and Fe/Al_2O_3Fe/Al_2O_3 tunnel barrier contacts for electrical spin injection into GaAs. Applied Physics Letters. 2004;**84**:4334. DOI: 10.1063/1.1758305

[4] Li CH, Kioseoglou G, van't Erve OMJ, Hanbicki AT, Jonker BT, Mallory R, et al. Spin injection across (110) interfaces: Fe/GaAs(110)Fe/GaAs(110) spin-light-emitting diodes. Applied Physics Letters. 2004;**85**:1544. DOI: 10.1063/1.1810534

[5] Olive Mendez S, Le Thanh V, Ranguis A, Derrien J. Growth of magnetic materials and structures on Si(0 0 1) substrates using Co_2Si as a template layer. Applied Surface Science. 2008;**254**:6040. DOI: 10.1016/j.apsusc.2008.02.193

[6] Olive-Mendez SF, Spiesser A, Michez LA, Le Thanh V, Glachant A, Derrien J, et al. Epitaxial growth of Mn_5Ge_3/Ge(111) heterostructures for spin injection. Thin Solid Films. 2008;**517**:191. DOI: 10.1016/j.tsf.2008.08.090

[7] Matsukura F, Ohno H, Shen A, Sugawara Y. Transport properties and origin of ferromagnetism in (Ga,Mn)As. Physical Review B. 1998;**57**:R2037. DOI: 10.1103/PhysRevB.57.R2037

[8] Jungwirth T, Wang KY, Masek J, Edmonds KW, Konig J, Sinova J, et al. Prospects for high temperature ferromagnetism in (Ga,Mn) As semiconductors. Physical Review B. 2005;**72**:165204. DOI: 10.1103/PhysRevB.72.165204

[9] Zutic I, Fabian J, Erwin SC. Spin injection and detection in silicon. Physical Review Letters. 2006;**97**:026602. DOI: 10.1103/PhysRevLett.97.026602

[10] Zhang FM, Liu XC, Gao J, Wu XS, Du YW, Zhu H, et al. Investigation on the magnetic and electrical properties of crystalline $Mn_{0.05}Si_{0.95}$ films. Applied Physics Letters. 2004;**85**:786. DOI: 10.1063/1.1775886

[11] Zhou S, Potzger K, Zhang G, Mucklich A, Eichhorn F, Schell N, et al. Structural and magnetic properties of Mn-implanted Si. Physical Review B. 2007;**75**:085203. DOI: 10.1103/PhysRevB.75.085203

[12] Ko V, Teo KL, Liew T, Chong TC, MacKenzie M, MacLaren I, et al. Origins of ferromagnetism in transition-metal doped Si. Journal of Applied Physics. 2008;**104**:033912. DOI: 10.1103/PhysRevB.78.045307

[13] Yabuuchi S, Kageshima H, Ohno Y, Nagase M, Fujiwara A, Ohta E. Origin of ferromagnetism of MnSi1.7 nanoparticles in Si: First-principles calculations. Physical Review B. 2008;**78**:045307. DOI: 10.1103/PhysRevB.78.045307

[14] Zwicker U, Jehn E, Schubert E. A study of the manganese-germanium system. Zeitschrift fuer Metallkunde. 1949;**40**:433

[15] Predel B. Ge-Mn (Germanium-Manganese). Landolt-Börnstein—Group IV Physical Chemistry, Volumn 5F. SpringerMaterials. 1996. DOI: 10.1007/10501684_1481

[16] Park YD, Hanbicki AT, Erwin SC, Hellberg CS, Sullivan JM, Mattson JE, et al. A group-IV ferromagnetic semiconductor: Mn_xGe_{1-x}. Science. 2002;**295**:651. DOI: 10.1126/science.1066348

[17] Ahlers S, Bougeard D, Sircar N, Abstreiter G, Trampert A, Opel M, et al. Magnetic and structural properties of Ge_xMn_{1-x} films: Precipitation of intermetallic nanomagnets. Physical Review B. 2006;**74**:214411. DOI: 10.1103/PhysRevB.74.214411

[18] Devillers T. Ferromagnetic phases of $Ge(1-x)Mn(x)$ for spintronics applications [PhD thesis]. Grenoble: Université Joseph Fourier; 2008

[19] Li H. Fabrication and characterization of nanostructured half metals and diluted magnetic semiconductors [PhD thesis] Singapore: National University of Singapore; 2006

[20] Fert A, Jaffrès H. Conditions for efficient spin injection from a ferromagnetic metal into a semiconductor. Physical Review B. 2001;**64**:184420. DOI: 10.1103/PhysRevB.64.184420

[21] Padova PD, Ayoub J-P, Berbezier I, Perfetti P, Quaresima C, Testa AM, et al. $Mn_{0.06}Ge_{0.94}$ diluted magnetic semiconductor epitaxially grown on Ge(001): Influence of Mn_5Ge_3 nanoscopic clusters on the electronic and magnetic properties. Physical Review B. 2008;**77**:045203. DOI: 10.1103/PhysRevB.77.045203

[22] Morresi L, Ayoub J, Pinto N, Ficcadenti M, Murri R, Ronda A, et al. Formation of Mn_5Ge_3 nanoclusters in highly diluted Mn_xGe_{1-x} alloys.

Materials Science in Semiconductor Processing. 2006;**9**:836. DOI: 10.1016/j.mssp.2006.08.056

[23] Bihler C, Jaeger C, Vallaitis T, Gjukic M, Brandt MS, Pippel E, et al. Structural and magnetic properties of Mn_5Ge_3 clusters in a dilute magnetic germanium matrix. Applied Physics Letters. 2006;**88**:112506. DOI: 10.1063/1.2185448

[24] Park YD, Wilson A, Hanbicki AT, Mattson JE, Ambrose T, Spanos G, et al. Magnetoresistance of Mn:Ge ferromagnetic nanoclusters in a diluted magnetic semiconductor matrix. Applied Physics Letters. 2001;**78**:2739. DOI: 10.1063/1.1369151

[25] Jamet M, Barski A, Devillers T, Poydenot V, Dujardin R, Bayle-Guillemaud P, et al. High-curie-temperature ferromagnetism in self-organized $Ge_{1-x}Mn_x$ nanocolumns. Nature Materials. 2006;**5**:653. DOI: 10.1038/nmat1686

[26] Olive Mendez SF, Petit M, Ranguis A, Le Thanh V, Michez L-A. From the very first stages of Mn deposition on Ge(001) to phase segregation. Crystal Growth & Design. 2018;**18**:5124-5129. DOI: 10.1021/acs.cgd.8b00558

[27] Alvídrez-Lechuga A, Holguín JT, Solís-Canto Ó, Santillán-Rodríguez CR, Matutes-Aquino JA, Olive-Méndez SF. Surface-interface quality of Mn_5Ge_3 thin films on Ge(001): Reactive deposition epitaxy vs. solid phase epitaxy. Microscopy and Microanalysis. 2018;**24**:1622. DOI: 10.1017/S1431927618008590

[28] Schütz MK, Petit M, Michez L, Ranguis A, Monier G, Robert Goumet C, et al. Thiol-functionalization of Mn_5Ge_3 thin films. Applied Surface Science. 2018;**451**:191-197. DOI: 10.1016/j.apsusc.2018.04.231

[29] Pinto N, Morresi L, Ficcadenti M, Murri R, D'Orazio F, Lucari F, et al. Magnetic and electronic transport percolation in epitaxial $Ge_{1-x}Mn_x$ films. Physical Review B. 2005;**72**:165203. DOI: 10.1103/PhysRevB.72.165203

[30] Tsui F, He L, Ma L, Tkachuk A, Chu YS, Nakajima K, et al. Novel germanium-based magnetic semiconductors. Physical Review Letters. 2003;**91**:177203. DOI: 10.1103/PhysRevLett.91.177203

[31] Michez L, Spiesser A, Petit M, Bertaina S, Jacquot JF. Magnetic reversal in Mn_5Ge_3 thin films: An extensive study. Journal of Physics: Condensed Matter. 2015;**27**(26):266001. DOI: 10.1088/0953-8984/27/26/266001

[32] Li AP, Wendelken JF, Shen J, Feldman LC, Thompson JR, Weitering HH. Magnetism in $Ge_{1-x}Mn_x$ semiconductors mediated by impurity band carriers. Physical Review B. 2005;**72**:195205

[33] Xie Y, Yuan Y, Mao W, Xu C, Hübner R, Grenzer J, et al. Epitaxial Mn_5Ge_3 (100) layer on Ge(100) substrates obtained by flash lamp annealing. Applied Physics Letters. 2018;**113**:222401. DOI: 10.1063/1.5057733

[34] Petit M, Boussadi A, Heresanu V, Ranguis A, Michez L. Step flow growth of Mn_5Ge_3 films on Ge(111) at room temperature. Applied Surface Science. 2019;**480**:529-536. DOI: 10.1016/j.apsusc.2019.01.164

[35] Kang JS, Kim G, Wi SC, Lee SS, Choi S, Cho S, et al. Spatial chemical inhomogeneity and local electronic structure of Mn-doped Ge ferromagnetic semiconductors. Physical Review Letters. 2005;**94**:147202. DOI: 10.1103/PhysRevLett.94.147202

[36] Picozzi S, Ottaviano L, Passacantando M, Profeta G, Continenza A, Priolo F, et al. X-ray absorption spectroscopy in Mn_xGe_{1-x} diluted magnetic semiconductor: Experiment and theory. Applied Physics Letters. 2005;**86**:062501. DOI: 10.1063/1.1861127

[37] Woodbury HH, Tyler WW. Properties of germanium doped with manganese. Physics Review. 1995;**100**:659. DOI: 10.1103/PhysRev.100.659

[38] Li AP, Zeng C, van Benthem K, Chisholm MF, Shen J, Rao SVSN, et al. Dopant segregation and giant magnetoresistance in manganese-doped germanium. Physical Review B. 2007;**75**:201201. DOI: 10.1103/PhysRevB.75.201201

[39] Devillers T, Jamet M, Barski A, Poydenot V, Bayle-Guillemaud P, Bellet-Amalric E, et al. Structure and magnetism of self-organized $Ge_{1-x}Mn_x$ nanocolumns on Ge(001). Physical Review B. 2007;**76**:205306. DOI: 10.1103/PhysRevB.76.205306

[40] Gambardella P, Claude L, Rusponi S, Franke KJ, Brune H, Raabe J, et al. Surface characterization of Mn_xGe_{1-x} and $Cr_yMn_xGe_{1-x-y}$ dilute magnetic semiconductors. Physical Review B. 2007;**75**:125211. DOI: 10.1103/PhysRevB.75.125211

[41] Biegger E, Staheli L, Fonin M, Rudiger U, Dedkov YS. Intrinsic ferromagnetism versus phase segregation in Mn-doped Ge. Journal of Applied Physics. 2007;**101**:103912. DOI: 10.1063/1.2718276

[42] Bougeard D, Ahlers S, Trampert A, Sircar N, Abstreiter G. Clustering in a precipitate-free GeMn magnetic semiconductor. Physical Review Letters. 2006;**97**:237202. DOI: 10.1103/PhysRevLett.97.237202

[43] Bougeard D, Sircar N, Ahlers S, Lang V, Abstreiter G, Trampert A, et al. $Ge_{1-x}Mn_x$ clusters: Central structural

and magnetic building blocks of nanoscale wire-like self-assembly in a magnetic semiconductor. Nano Letters. 2009;**9**:3743. DOI: 10.1021/nl901928f

[44] Zeng C, Zhang Z, van Benthem K, Chisholm MF, Weitering HH. Optimal doping control of magnetic semiconductors via subsurfactant epitaxy. Physical Review Letters. 2008;**100**:066101. DOI: 10.1103/PhysRevLett.100.066101

[45] Ahlers S, Stone PR, Sircar N, Arenholz E, Dubon OD, Bougeard D. Comparison of the magnetic properties of GeMn thin films through Mn L-edge x-ray absorption. Applied Physics Letters. 2009;**95**:151911. DOI: 10.1063/1.3232245

[46] Ahlers S, Bougeard D, Riedl H, Abstreiter G, Trampert A, Kipferl W, et al. Ferromagnetic Ge(Mn) nanostructures. Physica E: Low-dimensional Systems and Nanostructures. 2006;**32**:422. DOI: 10.1016/j.physe.2005.12.129

[47] Ahlers S. Magnetic and structural properties of $Ge_{1-x}Mn_x$ films: Precipitation of intermetallic nanomagnets. Physical Review B. 2006;**74**:214411. DOI: 10.1103/PhysRevB.74.2144

[48] Holy V. Diffuse x-ray scattering from inclusions in ferromagnetic $Ge_{1-x}Mn_x$ layers. Physical Review B. 2008;**78**:144401. DOI: 10.1103/PhysRevB.78.144401

[49] Sarney WL. Sample Preparation Procedure for TEM Imaging of Semiconductor Materials. Adelphi, Maryland: U.S. Army Research Laboratory; 2004. ARL-TR-3223; Corpus ID: 139408847

[50] Needels M, Payne MC, Joannopoulos JD. High-order reconstructions of the Ge(100) surface. Physical Review B. 1998;**38**(8):5543-5546. DOI: 10.1103/physrevb.38.5543

[51] Le TG. Direct structural evidences of epitaxial growth $Ge_{1-x}Mn_x$ nanocolumn bi-layers on Ge(001). Materials Sciences and Applications. 2015;**6**:533-538. DOI: 10.4236/msa.2015.66057

[52] Le TG, Dau M-T, Le Thanh V, Nam DNH, Petit M, Michez LA, et al. Growth competition between semiconducting $Ge_{1-x}Mn_x$ nanocolumns and metallic Mn_5Ge_3 clusters. Advances in Natural Sciences: Nanoscience and Nanotechnology. 2012;**3**:025007. DOI: 10.1088/2043-6262/3/2/025007

[53] Le TG, Nam DNH, Dau M-T, TKP L, Khiem NV, Le Thanh V, et al. The effects of Mn concentration on structural and magnetic properties of $Ge_{1-x}Mn_x$ diluted magnetic semiconductor. Journal of Physics: Conference Series. 2011;**292**:012012. DOI: 10.1088/1742-6596/292/1/012012

[54] Le TG, Le Thanh V, Michez L. Effect of carbon on structural and magnetic properties of $Ge_{1-x}Mn_xGe_{1-x}Mn_x$ nanocolumns. Bulletin of Materials Science. 2020;**43**:103. DOI: 10.1007/s12034-020-2082-z

[55] Le TG, Nuyen MA. Chemical composition of high-T_C $Ge_{1-x}Mn_x$ nanocolumns grown on Ge(001) substrates. Communications in Physics. 2014;**24**(2):163-169

[56] Le TG, Dau MT. Vertical self-organization of $Ge_{1-x}Mn_x$ nanocolumn multilayers grown on Ge(001) substrates. Modern Physics Letters B. 2016;**30**:1650269. DOI: 10.1142/S0217984916502699

[57] Zhu W, Weitering HH, Wang EG, Kaxiras E, Zhang Z. Contrasting growth modes of Mn on Ge(100) and Ge(111) surfaces: Subsurface segregation versus intermixing. Physical Review Letters. 2004;**93**:126102. DOI: 10.1103/PhysRevLett.93.126102

Chapter 4

Patchy Nanoparticle Synthesis and Self-Assembly

Ahyoung Kim, Lehan Yao, Falon Kalutantirige, Shan Zhou and Qian Chen

Abstract

Biological building blocks (i.e., proteins) are encoded with the information of target structure into the chemical and morphological patches, guiding their assembly into the levels of functional structures that are crucial for living organisms. Learning from nature, researchers have been attracted to the artificial analogues, "patchy particles," which have controlled geometries of patches that serve as directional bonding sites. However, unlike the abundant studies of micron-scale patchy particles, which demonstrated complex assembly structures and unique behaviors attributed to the patches, research on patchy nanoparticles (NPs) has remained challenging. In the present chapter, we discuss the recent understandings on patchy NP design and synthesis strategies, and physical principles of their assembly behaviors, which are the main factors to program patchy NP self-assembly into target structures that cannot be achieved by conventional non-patched NPs. We further summarize the self-assembly of patchy NPs under external fields, in simulation, and in kinetically controlled assembly pathways, to show the structural richness patchy NPs bring. The patchy NP assembly is novel by their structures as well as the multicomponent features, and thus exhibits unique optical, chemical, and mechanical properties, potentially aiding applications in catalysts, photonic crystals, and metamaterials as well as fundamental nanoscience.

Keywords: patchy nanoparticle, colloidal synthesis, self-assembly

1. Introduction

1.1 Definition and types of patchy particles

Anisotropic particles have been one of the fundamental focuses of materials research since their introduction several decades ago [1–4]. In the search for the next breakthrough functional materials, tremendous efforts have been dedicated to surface anisotropic particles. Surface anisotropy has proven to be vital in biological systems; examples include the surface structure of virus capsid, which enables the spread of disease by self-assembly of viral particles in vivo [5], and globular proteins with patches, which act as recognition sites [6].

Starting from Janus particles with two dissimilar hemispheres, anisotropic particles with dual or multiple surface patches [7], i.e., patchy particles, have been introduced over the past decades. Patchy particles are defined as a class of particles with discrete patches that induce strongly anisotropic and highly directional interactions, which can facilitate their self-assembly into ordered structures [8]. The patches are described as distinct features with limited numbers on the surface of core particles [8], and their properties, location, and numbers can be well controlled during synthesis [9].

The concept of patchy particles has been investigated in the realms of "hard" patchy colloids, hard colloids with "soft" patches, and soft patchy micelles. Hard colloids refer to colloid particles that do not exhibit morphological changes in suspension [10], including metal, metal oxide, and silica particles along with cross-linked polymer particles. Such patchy colloids can be synthesized from the controlled clustering of several individual particles and inorganic patch growth on parent particles [11, 12]. When the patches are soft materials such as polymer or DNA, they give different self-assembly behavior compared with hard counterparts. A recent example of this type of patchy particles is the gold triangular nanoprism with polymer patches on the three tips [13]. Lastly, patchy micelles are made from block copolymers with an insoluble core and soluble patches which can be distinguished chemically or physically [14]. Patchy particles are of interest in self-assembly as they generate unique assembly structures. For example, gold nanorods (AuNRs) with polymer patches were observed to self-assemble into clusters, linear and branched chains [15–17]. Furthermore, triblock copolymers are shown to first assemble into patchy micelles and further into linear and network superstructures [18].

1.2 Applications of patchy particles

Gaining control over the interactions between particles through their patches enables the potential bottom-up fabrication of functional materials for drug delivery, electronics, photonics, and sensors [19, 20]. Colloidal patchy polymers with the ability to self-knot have been proposed as prospective drug delivery vehicles [19]. These colloidal patchy polymers have a lock-unlock mechanism controlled by the end monomers, enabling the transport and release of drug molecules. DNA-patched particles with directional interactions governed by thermal reversibility have also been introduced with potential applications in drug delivery, solid-state electronic devices, and photonic crystals [21]. The self-assembled bidirectional percolated network and lattice structures of patchy metal-dielectric particles under high-frequency alternating current (AC) electric fields have been demonstrated as a promising method to achieve bottom-up fabrication of devices [20]. The self-assembly of these particles can be further guided by quadrupolar and multipolar interactions, and it has potential applications in the fields of photonics and electronics.

Although many self-assembled patchy particles with potential applications are being routinely presented, the current advancement in the field is not enough to meet the complex needs of future technological applications [8]. For synthetic self-assembled patchy particles to serve advantageous purposes in the fields of medicine, electronics, and photonics, structural complexity and hierarchy should be achieved, as those in biological functional structures. However, there are a few challenges in achieving such complex high order structures from synthetic patchy nanoparticles: the difficulty to precisely synthesize particles that are highly monodispersed in patch size and position; to arrange particles into highly ordered large-scale structures by fine-tuning the interaction; and to "position" the kinetic assembly pathways on the single particle level.

1.3 Scope

Nanoparticles are of special interest to the patchy particle community as they serve as versatile building blocks encoded with directional interactions, which enable spontaneous assembly into exotic structures for novel functional nanomaterials [22]. Hence the primary scope of this chapter will be patchy nanoparticles. This chapter covers both the experimental aspect—fabrication and self-assembly of patchy nanoparticles—and the simulation studies of their assembly behavior. We will first discuss controlling the multidimensional design space of patchy nanoparticles, with special considerations on the impact of core and patch design, followed by presenting three routes of patchy nanoparticle synthesis. A review of interactions driving self-assembly, including van der Waals (vdW), electrostatic, hydrophobic, functional group-based, other entropy-related, and field-assisted interactions, is then presented, with simulation of patchy nanoparticle assembly as a complimentary tool to guide and explain experiments. Finally, the rational design of kinetic assembly pathways for achieving three-dimensional (3D) hierarchical lattice is discussed, as well as our vision in the future directions of patchy nanoparticle assembly.

2. Preparation of patchy nanoparticles

2.1 Design rules of core and patch

Patches are distinct domains of surface chemistry (enthalpic patch) or structure (entropic patch) on particles. The regional inhomogeneity of patchy nanoparticles encodes directional interaction between patches, which determines the favorable interparticle alignment. Consequently, although core nanoparticles still matter, "patchiness" can completely alter the interaction dynamics and govern the final superlattice structures. The decades of extensive studies on conventional nanoparticles without patches have revealed the factors determining interaction between nanoparticles and their packing behavior, including shape anisotropy [23, 24], crystal facets [25], and entropy of the system [26, 27]. In recent years, both in experiment and theory, researchers have been trying to identify the design parameters for the patches to direct assembly behaviors of patchy nanoparticles. Some of those parameters are size [28], composition [29], surface chemistry [30, 31], symmetry [32], number of patches per particle [33, 34], as well as the binding energy between the patches [35]. While understanding the key design parameters is complicated because the impact of each is often hard to decouple from one another, some well-established design rules for patchy particles are listed herein. First, unlike the spherical nanoparticles without patches, whose coordination number in lattice is driven by the maximum packing density, in the case of the patchy nanoparticles, the "valency" is determined by the number of the patches. While symmetrically placed 12 patches on nanoparticles leads to the formation of face-centered cubic (*fcc*) superlattice due to the quasi-isotropic net interaction, decreased number of patches can generate lattices with lower structural symmetry and even open lattices that are hard to achieve by conventional nanoparticles [36, 37]. Note that even with the same number of patches and the same core shape, the particles can assemble into different final lattices as the angular symmetry of the patches changes. Second, the relative size of the patches to that of the core nanoparticle alters the valency of the patches. In the simplest case, the particles with one patch can undergo self-limited assembly into dimer, trimer, and tetramer clusters, as the relative patch size decreases. For the particles with multiple patches, optimal patch width is a trade-off between structural selectivity and kinetic accessibility [38]. Thus, the patches should be small enough to energetically favor the desired clusters. Meanwhile, they also need to be sufficiently large to bypass the kinetic traps, which will prevent the large-scale ordering. One of the successful examples achieving such is Kagome lattice by Chen et al. [36]. Following theoretical studies have revealed that the large-scale Kagome lattice is formed due to the balance in interactions caused by the patch area-dependent coordination number and entropy [39].

2.2 Synthetic routes of patchy nanoparticles

Inhomogeneous surface chemistry or local structure on nanoparticles can be obtained by a range of approaches, including colloidal assembly, phase separation, and site-specific surface modification. Such methods are mostly in solution base, thus can be easily scaled up with proper engineering. A library of patchy nanoparticles with controllable geometry and chemistry achieved herein paves the way to bring unprecedented assembly structures that cannot be realized by conventional nanoparticles with homogeneous surface.

2.2.1 Controlled assembly of nanocolloids

The assembly of conventional nanoparticles without patches into small clusters is a facile way to synthesize patchy nanoparticles. The immediate advantage of this approach is that two or more pre-synthesized nanoparticles with desired qualities (e.g., size, shape, monodispersity, and crystallinity) can be incorporated into a single moiety [40–42]. However, precise control over cluster size and shape requires optimal assembly condition [40]. First, the suspension of colloidal particles is destabilized due to interparticle interactions that are strong enough to assemble nanoparticles into clusters. Then the assembled clusters are prevented from further aggregation aided by surface charge, steric hindrance, or low enough particle concentration causing diffusion-limited assembly process. Such clusterization in a controlled manner can be realized by various types of interactions, including vdW, electrostatic, depletion, chemical bonding, or geometrical confinement.

The morphologies of patchy particles formed by clusterization are often driven by minimization of free energy. Wagner et al. [43] demonstrated the preparation of clusters composed of 2–7 spherical polystyrene (PS) nanoparticles, by combining ultrasonication-induced miniemulsions with dense packing—minimizing the second moment of the mass distribution, originally developed by Manoharan et al. [41]. Ultrasonic emulsification provided large yields of various patchy particles with a narrow size distribution, including line segments, triangles, tetrahedra, and octahedra (**Figure 1a** and **b**). Note that in order to achieve desired geometries, the subtle strength of interaction between nanoparticles is required to avoid rapid assembly and to allow the particles to rearrange within the clusters. When it comes to clusterization of binary composites with opposite charges, the balance of two competing forces, attraction toward the center particle and repulsions between the outer particles, at given feeding ratio assists controllable clusterization [44].

In case of metal or semiconductor nanoparticles with similar lattice constants of the core and the patch materials, oriented attachment can aid the clusterization process. Xiong et al. [45] synthesized heterodimers composed of silver prisms and Ag_2S spheres. Patchy nanoparticles with controllable size and thickness can be easily achieved by varying the dimension of prism and sphere constituting the structure. The as-synthesized patchy nanoparticles have remarkable bactericidal activity under illumination of visible light, attributed to the Ag/Ag_2S junction formed via oriented attachment. Furthermore, nanoparticles without the lattice

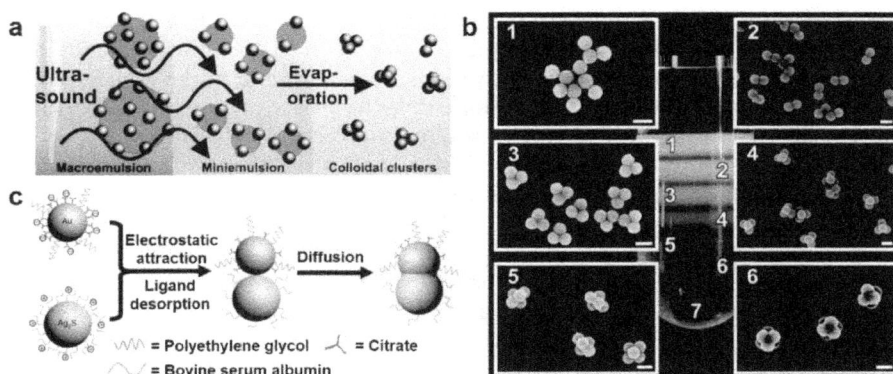

Figure 1.
(a) Schematics demonstrating cluster preparation. First, ultrasonication makes miniemulsion with toluene droplets (<2 μm in diameter) containing PS particles bound to the surface. Then the clusterization of the particles is driven during the evaporation of toluene. (b) The background photo shows a test tube containing a fractionized suspension of clusters differentiated by size after centrifugation in a density gradient. Inset scanning electron microscope (SEM) images show clusters obtained from six distinct bands: single particles (1), doublets (2), triplets (3), tetrahedrons (4), triangular dipyramids (5), and octahedrons (6). Fraction 7 contains clusters consisting of seven or more PS particles. Scale bars: 200 nm. Adapted from Ref. 44. Copyright 2008 American Chemical Society. (c) Mechanism of a two-step clustering and subsequent welding process making Au/Ag₂S patchy nanoparticles. Adapted from Ref. 46. Copyright 2019 Springer Nature.

match can cluster into well-defined composition by coulombic attraction. For example, using oppositely charged gold and chalcogenide nanoparticles, Huang et al. [46] demonstrated a method to synthesize patchy nanoparticles (**Figure 1c**). Their process involves a two-step mechanism of clustering and subsequent welding process, during which ligand desorption-induced conformal contact between nanoparticles allows atomic diffusion at the interface. Therefore, nanoparticles can assemble into patchy particles with desired composition, despite different capping agents, different solvent media, or large lattice mismatch between core and patch materials. Likewise, nanoparticle assembly via clusterization can provide a library of patchy nanoparticles with various patch composition and shape. However, the major drawback of this method is that patchy particles have low monodispersity in terms of the size and geometry of the clusters due to angular degeneracy. In order to improve the size dispersity, centrifugation or filtration can be used to separate the clusters of desired size among other types. Moreover, to achieve patchy nanoparticles with desired geometry, directional bonds of DNA wireframes can be used to guide tailored polyhedral shape, expanding the library of controlled assembly-derived patchy nanoparticles [47].

2.2.2 Phase separation-derived method

Phase separation is driven by the thermodynamic instability of a homogeneous mixture to lower the free energy of system upon the creation of two or more distinct phases. These mechanisms have been applied to polymer blends [48] and metallic glasses having components with different miscibility [49], in order to fabricate well-ordered or porous films that cannot be easily achieved by "top-down" nanolithography.

On the nanoparticle level, phase separation has also shown its value in preparing a variety of patchy nanoparticles with distinct surface chemistry, compositions, and shapes. For example, using the self-assembly induced phase separation of triblock terpolymer of polystyrene-*b*-polybutadiene-*b*-poly(methyl methacrylate), Muller and co-workers reported Janus, hamburger, clover, and football patchy nanoparticles [50, 51]. The different morphologies are attributed to the variations in the length of polymer blocks that selectively segregate upon reducing the solvent quality. Similarly, raspberry-like patchy nanoparticles have been demonstrated by using the emulsion solvent-evaporation process of polystyrene-block-poly(4-vinylpyridine) (PS-*b*-P4VP) solution. From these initially prepared patchy nanoparticles, Janus nanoparticles can be further derived by cross-linking the P4VP patches [52], followed by dissociating PS cores in good solvent. In addition, the composition and functionality of the particles are further extended, as metallic and oxide nanoparticles are preferentially grown within the P4VP domains upon precursor addition during phase separation. The same group also showed that patchy nanoparticles can possess hierarchical internal structures, by exploiting the phase separation of incompatible binary blends of amphiphilic block copolymer, PS-*b*-P4VP and homopolymer, poly(methyl methacrylate) (PMMA), in a confined 3D geometry (**Figure 2a** and **b**) [53]. Moreover, not only polymeric but also metallic patchy nanoparticles with internal hierarchy structure have been achieved by utilizing the alloy and phase-separated state of five metallic elements (Au, Ag, Co, Cu, and Ni) via polymer nanoreactor-mediated synthesis [54].

Phase separation can also be extended to the self-organization of ligands mixture on inorganic nanoparticle surface, to make chemically and topologically distinct patterns. For example, ligand layers composed of two types of thiol with different molecular lengths were observed to self-organize into striped patterns on gold nanospheres [55]. Chen and co-workers also reported polymer-patched gold nanospheres with controllable patch area and eccentricity from the core, as a function of ratio between two immiscible ligands (**Figure 2c–e**) [56, 57]. The controllable patch dimension is ascribed to the competition between hydrophilic and hydrophobic ligands attaching on the gold surface, followed by the adsorption of amphiphilic polymer only onto hydrophobic ligand-coated area. The phase separation of polymer solution into two phases, collapsed polymer and pure solvent, as reducing solvent quality has also been demonstrated on nanoparticles.

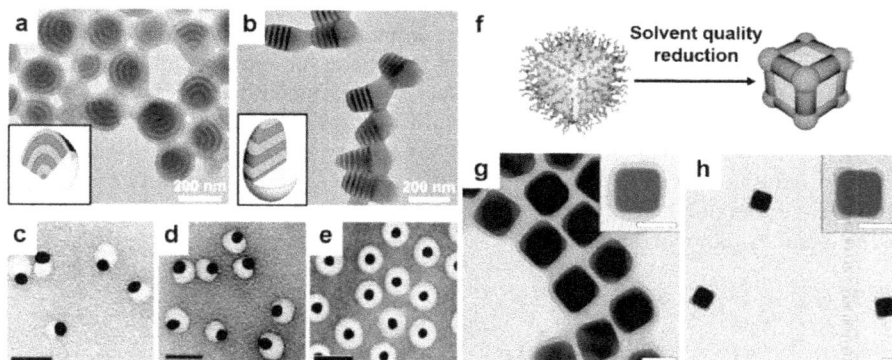

Figure 2.
(a and b) Transmission electron microscopy (TEM) images and schematics (insets) of PS-b-P4VP/PMMA patchy nanoparticles with hierarchical structure synthesized by emulsion solvent-evaporation process. Yellow, red, and green represent PMMA, PS, and P4VP, respectively. Adapted from Ref. 53. Copyright 2014 American Chemical Society. (c–e) Homocentric and eccentric Au@PSPAA nanoparticles depending on the hydrophobic and hydrophobic ligand ratio at 1:132 (c), 1:22 (d), and 1:0 (e). Adapted from Ref. 56. Copyright 2008 American Chemical Society. (f–h) Schematics of gold nanocube patterning with polymer patches and the corresponding TEM images. Surface-pinned micelle-like patches are formed upon reduction in solvent quality to homogeneously coated PS brushes on Au nanocube. (g) Nanocubes factionalized with PS forming a uniform polymer ligand shell in good solvent. (h) Patches formed on the vertices of nanocubes in poor solvent condition. Adapted from Ref. 59. Copyright 2017 American Chemical Society. Scale bars: 50 nm in (c–e, g and h).

Choueiri et al. [58] showed that in poor solvent, uniformly grafted PS brushes on nanoparticles can undergo thermodynamically driven segregation, forming surface-pinned micelle-like patches. The dimension, spatial distribution, and the number of patches are determined by the grafting density, solvent quality, relative size, and local surface curvature of the core particle (**Figure 2f–h**) [59]. Similarly, AuNRs with helicoidal PS patches are also achieved [60], by precisely controlling the molar ratio between two immiscible polymer brushes grafted on AuNRs. These polymer patches can be cross-linked to preserve the structure, or can go back to corona shell by increasing the solvent quality. Therefore, segregation-derived patch formation is a promising route to synthesize stimuli-responsive patchy nanoparticles, with morphological reversibility upon change of temperature, pH, salt concentration, etc.

2.2.3 Inhomogeneous surface modification-derived method

Surface modification is a powerful set of tools to tailor physical, chemical, or biological surface properties of nanoparticles required for a range of applications, such as abrasive resistance, biocompatibility, and colloidal stability in suspension. Surface engineering at the nanoscale is commonly done via bottom-up approach (e.g., overgrowth and ligand exchange), relying on chemical reaction of individual atoms and molecules to the nanoparticle surface. Thus, homogeneity of the modified nanoparticle surfaces depends on the thermodynamics and kinetics of the reactions, which are often challenging to be controlled site-specifically. In recent years, however, new knowledge acquired in heterogeneous nucleation and overgrowth, site-selective ligand exchange, and template-assisted materials deposition on nanoparticles, is paving ways to introduce patchiness on nanoparticles.

Seed-mediated nucleation and growth is a facile approach to synthesize multicomponent inorganic nanoparticles [61]. Among the possible morphologies of heterostructure nanoparticles, if the lattice strain induced by the mismatch is negligible, precursors form a shell that completely encapsulates the core. However, if the lattice strain is sufficiently high, the lattice relaxation induces the patchy crystal that partially deposited on nanoparticles [62–64], similar to the two dimensional (2D)-to-three dimensional transition in the Stranski-Krastanov model [65]. Utilizing this mechanism, various shapes of nanoparticles with crystal patches have been achieved, including rods, dumbbells and tetrapods [62, 66–69]. For example, Peng et al. demonstrated iron oxide nanoparticles with one to eight number of silver patches, via heterogeneous nucleation and growth approach [70]. In the synthetic process, first, the surface of iron nanoparticles is oxidized to generate amorphous iron oxide shells. Then silver patches overgrow on the shell, facilitated by the low enough interfacial energy between the amorphous iron oxide and crystalline silver. A similar mechanism has been adopted to synthesize gold-patched $MnO@SiO_2$ (**Figure 3a and b**) [71], and gold-patched anisotropic CdSe, PbSe, FePt, Cu_2O, and FePt–CdS nanocrystals [72]. Although the range of number and size of the patches can often be tuned by the reaction time or the concentration of precursor forming the patches [70], orthogonal control of such is not easy. It is because the event of heterogeneous nucleation and growth cannot be sharply separated by their nature. As a result, achieving highly monodisperse patchy nanoparticles in terms of the patch number (controlled by the number of heterogeneous nucleation events on one particle) and the size (controlled by growth) remains as a challenge. Moreover, if particles have multiple patches, controlling angular symmetry of the patches is not easy. Especially when the core particles are spherical, the position of patch-nucleation can be easily randomized, due to the lack of site-selectivity on homogeneous spherical nanoparticle surface.

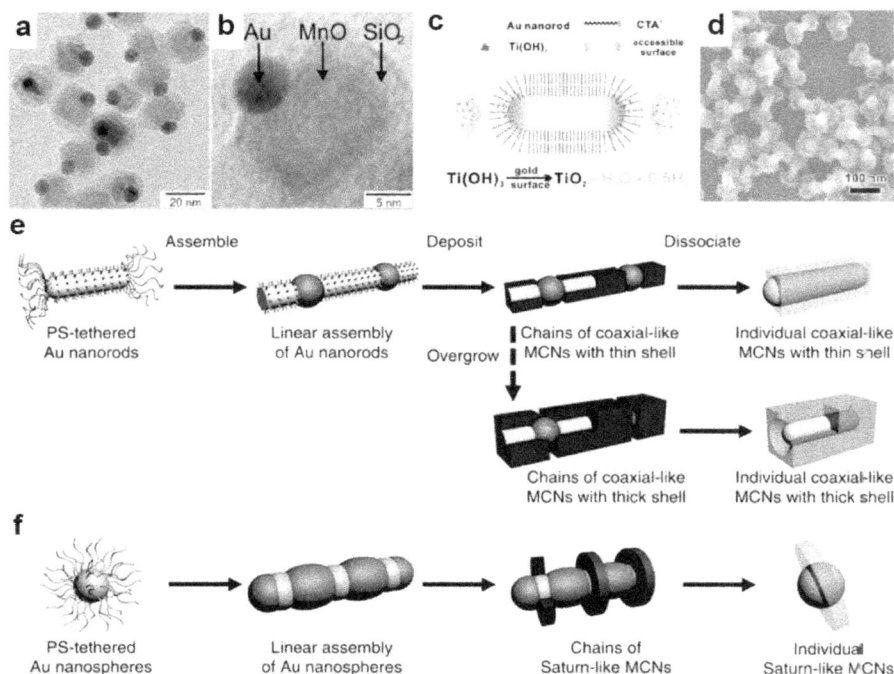

Figure 3.
(a) TEM image of gold patched MnO@SiO₂ nanoparticles. (b) TEM micrograph of a single gold patch MnO@SiO₂ nanoparticle showing a silica shell of ~3 nm thickness. Adapted from Ref. 71. Copyright 2014 American Chemical Society. (c) Representation of the less compact areas of CTAB layer on the AuNR, which facilitates the growth of TiO₂ to the tips of the AuNR. (d) SEM image of AuNR-TiO₂ dumbbells synthesized using CTAB capped AuNRs as a template. Adapted from Ref. 74. Copyright 2016 American Chemical Society. (e and f) Schematics of metal or metal oxide patch growth on (d) rod and (e) spherical nanoparticles using collapsed polymer shells to protect selected domains, making multicomponent colloidal nanostructures (MCNs). Adapted from Ref. 85. Copyright 2016 Springer Nature.

Anisotropic nanoparticles have intrinsic surface inhomogeneity, which can guide the site-selective ligand exchange and materials deposition. As an example, AuNRs have distinct surface curvature and crystal facets at the tip compared to the side. Due to this inhomogeneity, the common stabilizing surfactant, cetyl-trimethylammonium bromide (CTAB), binds less densely at the tip of AuNRs, exposing the tip area to be more susceptible for modification [73]. Tip-patched AuNRs with inorganic (e.g., TiO₂, SiO₂, etc.) and organic (e.g., PS, cysteine, etc.) patches have been achieved by site-selective overgrowth and ligand exchange at the tip (**Figure 3c and d**) [74–76]. Similarly, Kim et al. also reported nanoprisms tip-patched with polymer brushes, by exploiting the preferential chemisorption of 2-naphthalenethiol (2-NAT) on nanoprism tips, followed by polystyrene-*b*-poly-(acrylic acid) (PS-*b*-PAA) physisorption on ligands "island" via hydrophobic attraction [13]. The patch area and height can be orthogonally controlled as a function of 2-NAT and PS-*b*-PAA concentration, respectively. Various shapes of patchy prisms, including trefoil, T-shaped, and reuleaux triangle are demonstrated as a result of gradual patch expansion [13]. Likewise, patchy nanoparticle synthesis exploiting surface inhomogeneity provides site-selectively grown patches with decent monodispersity. However, this approach can be applicable only to the anisotropic nanoparticles with spatially distinct curvature or ligand distribution. Moreover, as the patch formation is governed by the local surface inhomogeneity, most of the patchy nanoparticles achieved herein have symmetric shapes as guided by the core particles' symmetry, posing difficulty in realizing asymmetrically patched anisotropic nanoparticles.

Template-assisted surface modification exploits immobilization of nanoparticles on solid surfaces, liquid-liquid interfaces, or liquid-gas interfaces [77–80], to expose only the part of nanoparticle surfaces to be accessible to the precursors dissolved in one phase. Using this approach, Janus nanoparticles with a library of compositions have been achieved, including metal-metal (e.g., FePt-Ag, Au-Ag), metal-oxide (e.g., Fe_3O_4-Ag), and polymer-metal [81–84]. Size of the patches can be kinetically tuned as a function of the precursor to seed-particle ratio and reaction time [80]. However, nanoparticles at the interfaces tend to tumble due to thermal and interfacial fluctuations, resulting in imprecise surface modification in terms of patch geometry and size. Another way to utilize template-assisted surface modification has been reported by Huang et al. (**Figure 3e** and **f**) [85]. In this method, selected domains of gold nanoparticle surface are protected by collapsed polymer shells, followed by the deposition of inorganic precursors on unprotected area into tubular metal or metal oxide patches. Similar concept has been used for silica-protected AuNRs to overgrow Ag, Pd and Pt metal patches on the exposed area [73], expanding the library of patchy nanoparticles with unprecedented shapes and properties.

3. Self-assembly of patchy nanoparticles

3.1 Experimental studies of patchy nanoparticle assembly

The bottom-up approach, exploiting the interactions between primary building units to assemble superstructures, has advantages of high throughput and energy efficiency, and low cost for nanofabrication [51, 86]. Beyond the conventional nanoscopic building blocks commonly making closely packed lattices, patchy nanoparticles have advantages in bringing an exotic library of programmable structures with tailored physicochemical properties ascribed to the directional interactions encoded via patchiness. As extensively studied in micron-scale patchy colloids systems [87], vdW and electrostatic interaction, hydrophobic attraction, chemical bonding, entropy-originated interactions, and field-assisted assemblies can all be utilized to induce patchy nanoparticle self-assembly. Nonetheless, due to two to three orders of magnitude difference in scale, nanoscopic self-assembly compared to that at micron-scale has intrinsic differences worth our attention. To list some, interaction between nanoparticles is long-range relative to their size, allowing recognition and specific alignment of nanoparticles even when they are physically apart [88, 89]. Moreover, unlike those of micron-scale particles, nanoparticle interactions are more drastically controlled by particles' local morphology details (e.g., truncation at the nanoprism tip), possibly due to the long-range effect and facet-dependent density of capping ligands. Lastly, quantitative understanding of nanoparticle interaction is still in its infancy, as often the interaction is non-additive, and theories developed for micron-scale colloids is not applicable, or difficult to be tested in situ [90]. Despite such complexity, the listed distinctive features of nanoscale self-assembly bring huge opportunities in designing self-assembly structures and understanding self-organization phenomena found in nature and biology. Acknowledging such opportunities, here we summarize self-assembly strategies stemmed from different types of interactions that have been successfully exploited for patchy nanoparticle.

3.1.1 DLVO interaction

vdW force, originated from the instantaneous interaction of permanent or transient dipoles between the atoms and molecules, is one of the major forces causing random aggregation of nanoparticles and uncontrolled assembly. To prevent such instability of nanocolloids, electrostatic repulsion is often introduced by grafting charged ligands onto the surface of nanoparticles. This electrostatic repulsion can be easily tuned by varying the ligand density and the charge screening effect, which are determined by the ionic strength of the nanoparticle suspension [91]. The interplay between vdW attraction and screened electrostatic forces can collapse into a Derjaguin-Landau-Verwey-Overbeek (DLVO) interaction energy profile [92], which controls the overall interaction of nanoparticles and final self-assembly structures [87]. Although such interaction is nondirectional in general, in case of patchy nanoparticles, the distinct domains of patched and non-patched regions with different physicochemical properties introduce site-specific DLVO interactions, allowing direction-dependent attraction or repulsion over the surface.

As an example, the gold patched organosilica nanoparticles covered with CTAB molecules can reversibly assemble into well-controlled clusters with "hot-spots" consisting of gold patches in physical contact [29], by controlling the DLVO interaction between gold patches (**Figure 4a–c**). The reversible patch-to-patch assembly is ascribed to the dominating gold-gold vdW attraction as the positively charged CTAB molecules on gold patches are washed out in ethanol. Dimers and trimers are obtained by tuning the steric hindrance as a function of the relative patch size. Compared to the self-assembly driven by covalent bonds of chemical linkers, exploiting DLVO interaction has advantages in that it is fast, reversible, often low cost, and requires less intensive synthesis efforts [29]. Triangular gold nanoprisms tip-patched with negatively charged polymers also show controlled self-assembly into twisted dimers, star and slanting-diamond (**Figure 4d** and **e**) [13]. By increasing the ionic strength of the patchy nanoprism suspension, the electrostatic repulsion between the particles is screened, and the exposed vdW force between the non-patched gold prism surfaces causes assembly into dimers. The tip-patched nanoprisms could assemble into large-scale 2D lattice if the directional repulsion introduced by patches is controlled to render specific angle between the particles. Zhu et al. [93] achieved the *fcc* superlattices from near-spherical quantum dots patched with gold satellites (**Figure 4f** and **j**), in which the orientations of individual nanoparticles are aligned. Using both small-angle X-ray scattering (SAXS, **Figure 4g** and **k**) and wide-angle electron diffraction (**Figure 4h, i, l** and **m**), they showed that by varying the ligand thickness; the degree of alignment can be controlled through the distance-dependent vdW attraction. It is expected that upon decreasing the number of gold patches, the particles could form into the 3D structures with lower coordination numbers, including body-centered-cubic and diamond lattices. Achieving these exotic self-assembly structures by controlling directional DLVO interactions of patchy particles can be universal, and thus many research interests are focused on experimental efforts to improve the purity of nanoparticles and to precisely control the interaction [94].

3.1.2 Hydrophobic attraction

Hydrophobic attraction originates from the entropy gain of the system when hydrophobes in aqueous solution aggregate to minimize their contact with water molecules, which otherwise make ordered cage-like structure on hydrophobic surfaces [95]. As already exploited by nature during protein folding into functional

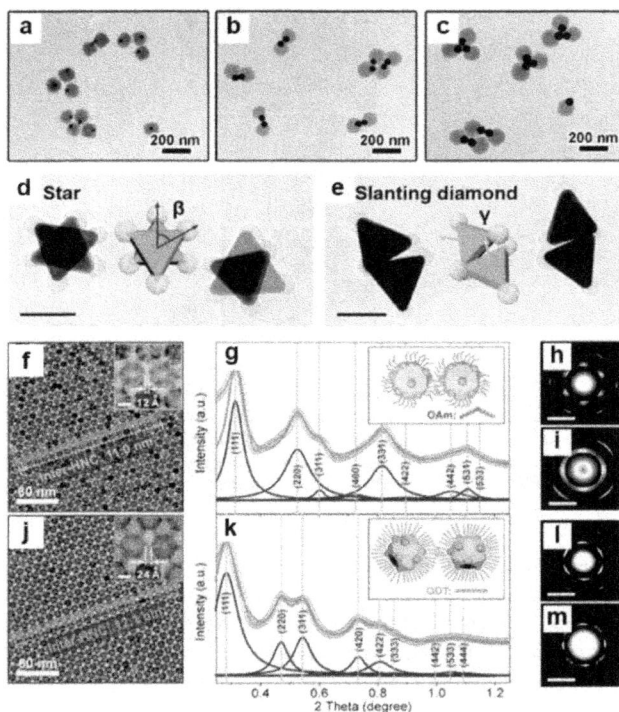

Figure 4.
(a–c) TEM characterization of assembled Au patched organosilica nanoparticles with different steric hindrance. The size of Au patches is (a) 24, (b) 44, and (c) 62 nm, respectively. The size of organosilica nanoparticle is about 120 nm. Adapted from Ref. [29]. Copyright 2016 American Chemical Society. (d and e) Self-assembly of tip-patched Au nanoprisms into twisted star (d) and slanting diamond (e) structures. Adapted from Ref. [13]. Copyright 2019 American Chemical Society. (f–m) Control of patchy nanoparticle orientations via ligand layer thickness. (f and j) TEM images of the nanoparticle superlattice (SL) viewed along the close-packed $(111)_{SL}$ zone axis, assembled from nanoparticle with different surface ligand density. Insets: Zoomed-in TEM images of the superlattice. (g and k) SAXS patterns of the corresponding superlattices. (h, i, l, and m) Wide-angle electron diffraction patterns along $(111)_{SL}$ (h and l) and $(001)_{SL}$ (i and m) of the assembled superlattices. Adapted from Ref. [93]. Copyright 2019 American Chemical Society. Scale bars: 50 nm in (d and e); 5 nm in (f and j insets); 2 nm^{-1} in (h, i, l, and m).

structures and phospholipid assembly into bilayer membranes, hydrophobic attraction can also be utilized to achieve controlled assembly structure by introducing the hydrophobic patches on nanoparticles with hydrophilic surface, or vice versa. By grafting PS on both ends of the AuNRs, Nie et al. achieved tip-to-tip, linear, or ring self-assembly structures ascribed to the directional hydrophobic attraction between the PS on AuNR tips [15]. The following study further shows that the solvent quality and the molecular weight of PS play an important role in controlling the strength of hydrophobic attraction, and thus provide diverse structures, including short bundles and bundled chains (**Figure 5a–d**) [96]. The linear assembly of PS-grafted AuNRs resembles step growth polymerization. The kinetic study of the system shows that the aggregation number of the chains increases linearly with reaction time, which is a characteristic of reaction-controlled step growth polymerization [97]. Other studies also report that concepts adapted from conventional polymerization, such as copolymerization and chain stoppers can be implemented to make a library of assembly structures with controllable composition and dimension [98, 99]. Similar to the polymer-patched inorganic particles, soft micelle nanoparticles prepared from diblock or triblock copolymers also show the controllable assembly

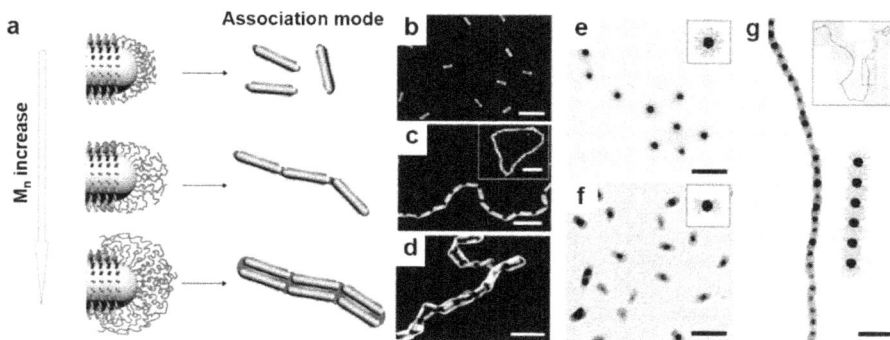

Figure 5.
(a) Left column: Schematics of PS with varying molecular weight on AuNRs. Central column: Schematics of different self-assembled structures of AuNRs. (b–d) SEM images of different self-assembled structures of AuNRs. Adapted from Ref. [96]. Copyright 2008 American Chemical Society. (e) Spherical micelles of polystyrene-poly(4-vinylpyridine) (PS-P4VP). (f) Patchy nanoparticle monomers of PS-P4VP. (g) A self-assembled polymer chain of PS-P4VP patchy nanoparticles. Adapted from Ref. [100] with permission from the Royal Society of Chemistry. Scale bars: 100 nm.

induced by solvophobic attraction between the patches (**Figure 5e–g**) [100, 101]. For example, upon gradual decrease of the solvent quality, polystyrene-block-polybutadiene-block-poly(methyl methacrylate) molecules undergo staged assemblies: micelle formation, dimerization of micelles into double-patched particles, followed by further assembly of patchy nanoparticles via step-growth polymerization and branching [51]. As a result, the intrinsically self-assembled soft patchy nanoparticles exhibit a complicated final assembly structures in 1D (chain) or 2D (network) [18]. Such hierarchical complexity induced by patchiness is promising for achieving abundant and sophisticated structures and yet to be further explored at nanoscale (further discussion in Section 3.3).

3.1.3 Formation of specific bonds between functional groups

Specific bonds between functional groups, including covalent bond and hydrogen bond (i.e., DNA base-pairing) on the site-selectively modified nanoparticles, can trigger programmable self-assembly, as encoded by functional groups [102]. A well-known example of the specific bond directed self-assembly is the biotin-streptavidin pair [103]. Herein, both tips of the AuNR are first patched with biotin disulfide. Then the tip-modified AuNRs gradually assemble into linear chain-like structures when streptavidins are added to covalently "link" two disulfide groups (**Figure 6a** and **b**). Another recent example utilizes the chemical reaction between amine group and carboxyl group, inducing the self-assembly between concavely patched silica nanoparticles and spherical silica nanoparticles [104]. The self-assembled structures are clusters with sp, sp^2, and sp^3 hybridized molecular structures (**Figure 6c–h**), which can be potentially used as building blocks for hierarchical self-assembly. Nanoparticle assemblies through DNA linkage have gained extensive research interests to encode directional bonding between the nanoparticles, after first reported by Mirkin and Alivisatos groups [105, 106]. Especially, the patch formation by regioselectively coated DNA linkers on the nanoparticle surface is a promising strategy to form programmable, unprecedented structures [30, 31, 107]. Xu et al. [30] synthesized gold nanospheres asymmetrically grafted with oligonucleotide DNA linkers, by utilizing magnetic microparticles as templates for the nanospheres to attach upon and undergo partial surface modification on the exposed region. These patchy nanospheres are co-assembled with nanoparticles

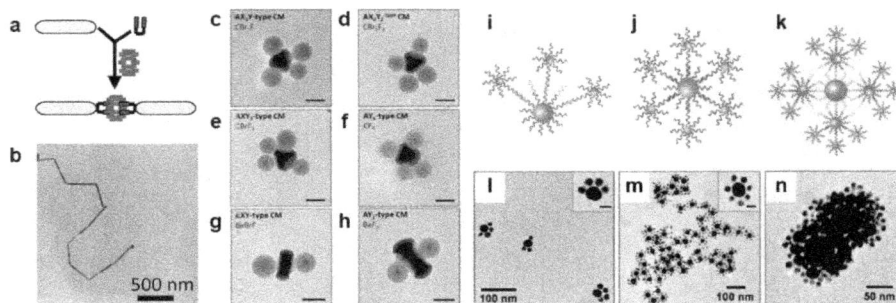

Figure 6.
(a) Schematics showing the assembly of AuNRs (golden ovals) by surface functionalization with the biotin disulfide (red), and subsequent addition of streptavidin (blue), to produce aggregates of AuNRs. (b) TEM image of assembled AuNRs, surface-derivatized with either biotin disulfide, after addition of streptavidin. Adapted from Ref. [103]. Copyright 2003 American Chemical Society. (c–f) TEM images of the colloid molecules assembled from mixing particles with four aminated dimples with 100 nm silica nanospheres and 90 nm core–shell nanoparticles in different ratio. (g and h) TEM images of the colloid molecules assembled from mixing particles with two aminated dimples with 100 nm silica nanospheres and 90 nm core–shell nanoparticles in different ratio. Adapted from Ref. [104]. Copyright 2018 Beilstein-Institut. (i–n) Self-assembly of partially functionalized gold nanoparticles into (i and l) cat paw, (j and m) satellite, and (k and n) dendrimer-like structures. Adapted from Ref. [30]. Copyright 2006 American Chemical Society. Scale bars: 100 nm in (c–h); 20 nm in insets of (l and m).

uniformly coated with DNAs into cat paw-, satellite-, and dendrimer-like structures (**Figure 6i–n**). Others also use isotropic gold nanoparticles embedded in highly anisotropic DNA origami to achieve self-assembly structures controllable from zero dimensional clusters to 3D diamond family superlattices [31, 107]. Chen and co-workers [56, 57] and Chen et al. [108] explore the self-assembly of a variety of gold nanoparticles (e.g., nanorods, nanoprisms, and nanocubes) site-specifically patched with single-stranded DNA (ssDNA) molecules. The nanoparticles are selectively covered by polymer shells, and ssDNAs are attached onto remnant area of nanoparticle surface (e.g., tip, edge, and face). As the ssDNAs on patches hybridize, the nanoparticles self-assemble into various structures, including dumbbell-, exclamation mark-, pearl necklace-, starfish-, and snowman-like clusters. The self-assembly of DNA-patched nanoparticles into large-scale lattices are yet to be achieved, aided by developing strategies: site-specifically functionalizing nanoparticles and assembling nanoparticles homogeneously coated by DNAs.

3.1.4 Other interactions of an entropic origin

Entropy effects other than hydrophobic interactions on nanoparticle self-assembly are commonly originated from depletion attraction or from the densest packing of particles in suspension of maximum translational entropy [87]. The depletion attraction is triggered when depletants of small sizes exist in the colloidal suspension [109], which has excluded volume in the vicinity of the colloids. The total entropy of the system increases as the particles come into contact and cause the overlapped excluded volume to increase and allow the depletants to explore more space [110]. The phenomenon can be interpreted as when depletants are absent in the space between two particles, the lower osmotic pressure of the pure solvent in this volume pushes particles toward each other [94]. Depletion attraction has been successfully exploited to trigger self-assembly of micro-sized concave patchy colloids as "lock-and-key" and non-patchy anisotropic nanoparticles with large flat facets (**Figure 7a–d**), in the presence of surfactant micelles or non-adsorbing water-soluble polymers [111–114]. The entropy-induced packing of shape-anisotropic

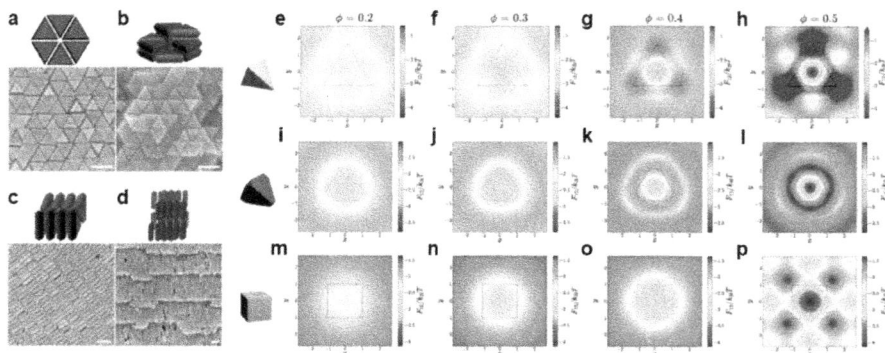

Figure 7.
(a–d) Schematics and SEM images of different depletion attraction-driven nanoprism self-assembly structures: (a) single-layer p-honeycomb, (b) multilayer p-honeycomb, (c) single-layer i-honeycomb, and (d) multilayer i-honeycomb. Adapted from Ref. [114]. Copyright 2017 American Chemical Society. (e–p) Potential wells by taking slices of the potential of mean force and torque (computed from the frequency histogram of the relative Cartesian coordinates of pairs of particles in Monte Carlo simulations of monodisperse hard particles) parallel to the faces of a (e–h) tetrahedron, (i–l) tetrahedrally faceted sphere, and (m–p) cube at various packing fractions ϕ = 0.2, 0.3, 0.4, and 0.5 as indicated at the head of each column. Adapted from Ref. [115]. Copyright 2014 National Academy of Sciences. Scale bars: 100 nm in (a–d).

nanoparticles has been extensively studied on a theoretical level [115–117]. The driving force of assembly, herein, is often referred as "entropic patchiness" [118, 119] due to the preferential interaction between the planar face over curved surface regions (**Figure 7e–p**). This preference has been rationalized as the emergence of "directional entropic forces" [24, 120]. In experimental systems, entropic patchiness is frequently employed to achieve the self-assembled superlattice from shape-anisotropic nanoparticles. Murray and co-workers demonstrated that $NaYF_4$ nanoparticles with different geometries, including spheres, rods, hexagonal prisms, and plates, self-assemble into densely packed lattices as entropically-directed [121]. They further showed from experimental and computational investigations that hexagonal nanoplates self-assemble into long-range ordered tiltings, ascribed to the balance between shape-induced entropic and edge-specific enthalpic interactions [122]. Moreover, entropic patchiness-driven self-assembly strategy is not only limited to single-component but can be extended to binary-components with compatibility in size and shape [123, 124]. For example, Ye et al. [123, 124] demonstrated that the relative size of AuNRs and gold nanospheres affects the entropy of the system and triggers the co-assembly of nanoparticles into three different types of phases, including lamellar structure with disordered spheres and AB_2-type binary superlattice. Optimum design rules for entropic patchy nanoparticle combinations are yet to be further explored both in experiment and computation, and promise a richer library of superlattices and phase behaviors beyond those from a single component.

3.1.5 Field-driven assembly

Patchy particles, when appropriately designed, have programmable direction and magnitude of interparticle interaction controlled by types of external field (e.g., electric and magnetic), field properties [125], as well as the composition and shape of the particles. As a result, patchy particles can align to the field, or transform the input energy into mechanical energy to exhibit out-of-equilibrium dynamic behaviors such as self-propulsion or rotation. Moreover, the magnitude of interparticle interaction can be altered from one to three orders of k_BT, as a function of the strength of field and frequency, providing an effective "switch" to reversibly control the interaction and assembly.

AC electric field is commonly used to alter the particle interactions while effectively suppressing the electrolysis of the liquid in comparison to the use of direct current electric field. Compared to non-patchy particles, patchy particles could exhibit richer assembly structures due to their unique shapes and special arrangements of induced dipole moments. For example, under AC electric field, spherical or ellipsoidal colloids with homogeneous surface charge assemble into a face-centered ABC layer packing [126]. By contrast, patchy particles with negatively charged patches under AC electric field can form 1D, 2D, 3D, and even double-helix structures (**Figure 8a**) [127]. Such helical structure stems from the charge redistribution on patchy particle surface where patches align to maximize the attraction along the structural axis. Another example shows that particles with a protruding patch assemble into various chiral clusters under AC electric field (**Figure 8b**) [128]. The diverse structures where patches point toward the chiral center are attributed to the induced dipolar interactions tunable by field frequency and patch size. The chirality of clusters induces imbalanced hydrodynamic flow, leading to the rotation of clusters in the direction opposite to their handedness. Chiral clusters with similar behavior are also co-assembled from self-propelling metal-dielectric patchy particles in combination with non-patchy particles, under AC electric field [129]. Likewise, if the patchy particles have compositions with distinct dielectric permittivity (e.g., titanium/silica), the anisotropy in the magnitude and in frequency-dependent responses trigger various modes of self-propelled behaviors [130]. For metal-dielectric patchy particles subjected to the AC electric field of different frequencies, dynamic collective structures of swarms, chains, and clusters have been achieved. Moreover, if triangular Au-patches are formed on PS particles, the particles were aligned and propelled following helical trajectories under AC (**Figure 8c–e**) [131]. The structural and behavioral similarity in these studies suggests the existence of governing rules when it comes to assemble patchy particles under electric fields, possibly due to the overwhelmingly large dipole interactions relative to thermal fluctuations.

Magnetic responses of patchy particles usually require paramagnetic or ferromagnetic compositions. Permanent magnetic moment in ferromagnetic patchy particles often allows the assembled structures to be preserved even after the external magnetic field is turned off. Granick group and the others have investigated the rich superstructures with distinct crystalline symmetries formed by ferromagnetic patchy particles [132, 133]. For example, under the high-frequency magnetic field, nickel-deposited silica patchy particles pair into dimers that subsequently assemble into reconfigurable hierarchical structures, including zigzag chains, square, and hexagonal lattices [125]. Other structures, such as chains and layered structures (**Figure 8f–i**), and microtubes have been achieved by applying static or precessing magnetic fields [134–136]. At nanoscale, under static magnetic field, Fe_3O_4-Ag heterodimeric, ferromagnetic patchy nanoparticles can be assembled into densely packed helical superstructures [137]. As the size of silver patch increases, gradual structural transition from helix to belt is observed, resulting from the balance among magnetic dipole-dipole interactions, vdW attraction, Zeeman coupling, and entropic forces.

Comparing to the phenomenal amount of in-depth studies on dynamic assembly of micron-scale particles, there are only a few studies on the field-driven assembly of patchy nanoparticles. It is mainly caused by the technical difficulties posed in lack of imaging tools for nanoscale structures in situ [138]. Nonetheless, dynamic assembly of patchy nanoparticles has a huge potential in offering unanticipated structures and exotic properties for the future nanomaterials. The recent technical

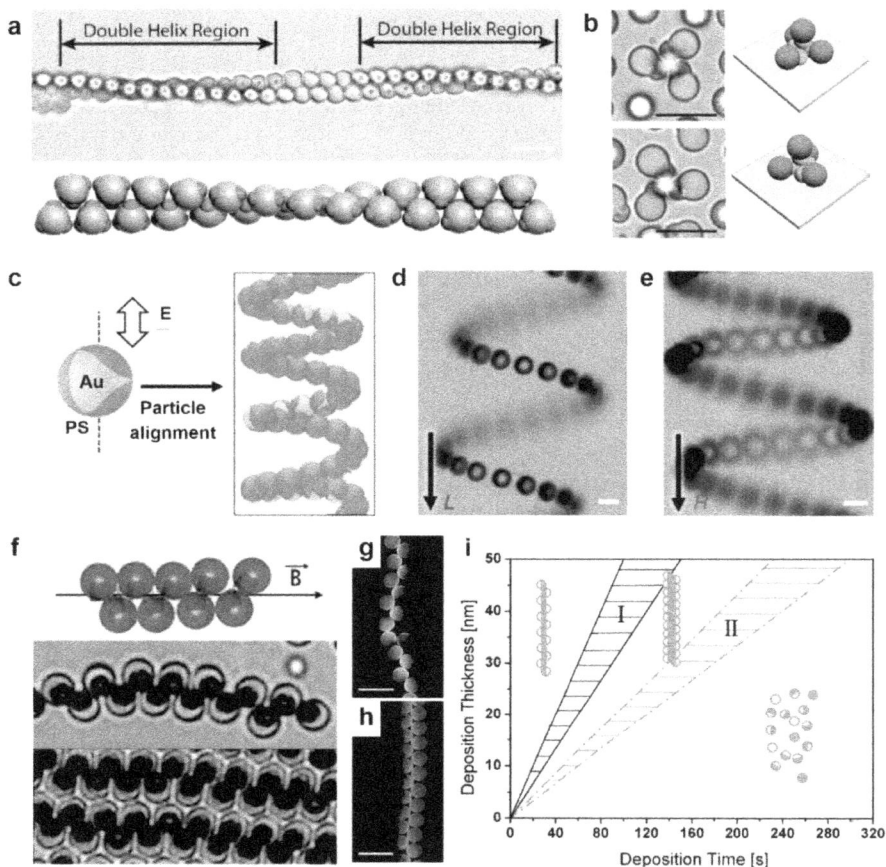

Figure 8.
(a) Optical microscopy (OM) image and schematics of double helix regions formed from tripatch particles under AC electric field. The double helix regions in the two paired chains are denoted by red and blue dots. Adapted from Ref. [127]. Copyright 2015 American Chemical Society. (b) OM images and schematics of chiral tetramers assembled from spherical particles with protruding patch under AC electric fields. Adapted from Ref. [128]. Copyright 2015 National Academy of Sciences. (c) Schematic and (d and e) OM images of Au-patched PS particles alignment following helical trajectories under AC-electric field, denoted as E. The arrows represent the direction of helical motion of patchy particles. L and R denote left and right-handed helical trajectories. Adapted from Ref. [131]. Copyright 2019 Springer Nature. (f) Schematics and OM snapshots of spherical particles with a magnetic patch assembling into linear chains and layered structures under external magnetic field, denoted as B. Adapted from Ref. [135]. Copyright 2012 American Chemical Society. (g and h) SEM images of particle assemblies into (g) staggered chain and (h) double chain after drying the particle suspension under static magnetic field. Brighter areas on the particles are iron oxide caps, and darker regions are unmodified PS surface. (i) Assembly behavior diagram as a function of iron oxide deposition thickness and time indicating regions of staggered chain, double chain, and no assembly behavior. Regimes denoted as I and II refer to transition behavior, where deposition parameters lead to particles exhibiting either of the adjacent assembly behaviors. Adapted from Ref. [134]. Copyright 2012 American Chemical Society. Scale bars: 5 μm in (a, b, d, e, g, and h).

development in liquid-phase transmission electron microscopy [139], which enables the in situ imaging of the nanoparticle suspension in real time and real space at the single particle level, is a promising and powerful tool to overcome the current difficulties. Applying the electric fields and decoupling the electron beam effect from the generic behavior of the particles [88, 140], or utilizing electron beam itself as a handle to manipulate the behavior of nanoparticles [141] is gaining increasing research interest. Assembly mechanism and various structures observed both from in situ micron-scale patchy particles and ex situ patchy nanoparticles can be important guidelines for the in situ studies at nanoscale in the future.

3.2 Simulation of self-assembly behaviors of patchy nanoparticles

For the past two decades, simulation of patchy particles has been extensively developed using simplified models to understand complicated assembly systems [142], including protein folding [143], viral capsid formation [144, 145], and even phase transition of atoms and molecules [146]. In the field of synthetic patchy nanoparticles, simulation aids understanding of experimental observations. For example, Lu et al. [147] experimentally showed nanocubes whose corners have ssDNA patches pack into unique lattices with zigzag arrangement. By adapting a scaling theory and Monte Carlo simulation for the ssDNA grafting on cubes, they predicted the shape of the corona morphology transforming from face-preferred to corner-preferred, as ssDNA length increases. They further developed the perturbation theory and potential mean force calculation for the set of patchy nanocubes to reveal the crucial role of concave or convex patches in controlling the angular arrangement of particles in the lattice. The fidelity of the simulation is further confirmed by comparing to the SAXS measurement of the lattices. Walker et al. [148] experimentally showed 3D open-lattices assembled from gold nanodumbbells with chemical patches on the neck of the particles. From finite-element calculations, they confirmed the most favorable alignment of two nanodumbbells is the cross-stack of the particles by their patches, consistent with the unit structure of the lattice. Zhu et al. [93] reported quantum dots decorated with gold patches and assembled them into close-packed structures. The molecular dynamics simulation not only reproduced their superlattices but also suggested a means of tuning the orientational alignment of the building blocks in the superlattice by varying distance-depended vdW force between the gold patches. Complementarily, in their experiments, the interparticle distance is controlled by the ligand thickness, and the orientational alignment change is confirmed by wide-angle electron diffraction.

Moreover, simulation often has been regarded as the only method to systematically study impacts of each geometric and interaction parameter constituting patchy particles, especially at the nanoscale where direct imaging of assembly behaviors has remained as a challenge. Systematic variation and decoupling of geometric factors and interactions can be achieved by simulation to guide experiments. Smallenburg and Sciortino showed patchy particles can have stable liquid phase that does not transform to gel structures even at the zero-temperature limit [146]. Their study suggests the flexibility of bonds and limited valence, which can be controlled by the patch number and size are the crucial factors in designing stable amorphous structures and glass-forming molecular networks. The same group also suggests through rational design of patch shape and symmetry that triblock patchy particles can selectively crystallize into tetrastack lattice with unique photonic properties, and even a colloidal clathrate-like structure [149]. Other structures including icosahedra, tetrahedra, square pyramids [8], helical structures [150], quasicrystals of dodecagonal symmetry [35, 151], and open lattices [152, 143] (**Figure 9a** and **b**) with high structural selectivity over other polymorphs are studied (**Figure 9c**) [153]. Together with recent studies in mechanical and functional properties of the exotic structures formed as bottom-up [154, 155], these theoretical studies have been paving ways to experimentally achieve novel functional nanostructures.

Lastly, simulation helps the experimental design of patchy nanoparticles to achieve targeted assemblies with desired properties via "inverse design" [117]. Inverse design is a powerful strategy to find crucial geometric parameters and physical interactions for patchy nanoparticles, especially to realize unprecedented

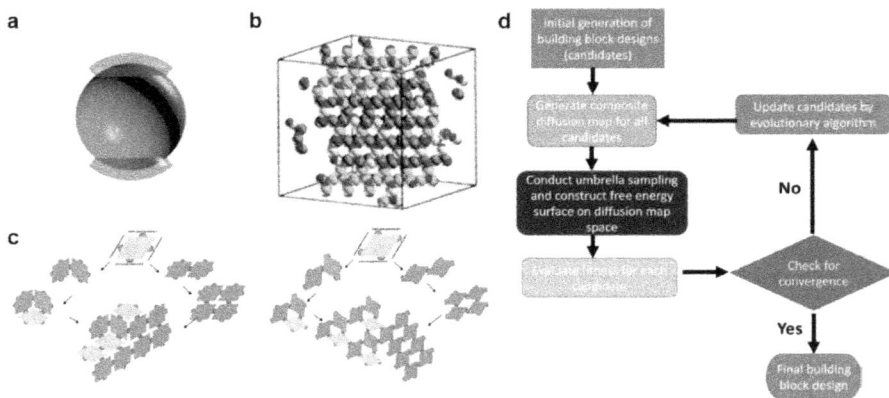

Figure 9.
(a and b) Molecular dynamics simulation snapshots of triblock patchy particles. (a) Model spherical particle has two attractive patches (blue shell). (b) Simulation snapshots of crystalline structure from triblock patchy particles (patches in blue). Particles are colored according to the local lattice environment: green and red particles are in a cubic and hexagonal structure, respectively, while yellow particles are in a mixed local environment. Adapted from Ref. [152]. Copyright 2016 American Chemical Society. (c) Schematics of close packed tilting to open lattice structure obtained from simulation of rhombi particles with differently located four patches. Adapted from Ref. [153]. Copyright 2019 American Chemical Society. (d) Flowchart of the landscape engineering inverse design procedure. Adapted from Ref. [156]. Copyright 2019 Royal Society of Chemistry.

3D structures. By engineering the free energy landscape via evolutionary algorithms, Ma and Ferguson discovered patchy particles that can hierarchically assemble into pyrochlore and cubic diamond lattices with complete photonic bandgaps (**Figure 9d**) [156]. Moreover, a generalizable inverse statistical mechanism approach is developed by Chen et al. [157] to provide a set of design rules for a collection of 2D crystals, including square, honeycomb, kagome, and parallelogrammic lattices. Along the similar direction, Whitelam reported a strategy of minimal positive design and identified distinct types of interaction between patchy particles required for achieving eight kinds of Archimedean tilting plane structures [158].

3.3 Design of kinetic pathways: To achieve 3D lattice via hierarchical assembly

Designing 3D open lattices with complicated structural details and low coordination number presents a challenge because conventional non-patchy particles tend to pack closely, instead of allowing for periodic arrays of holes [107]. Open and highly ordered structures have tailorable properties, including light weight, high porosity, low thermal conductivity, tailored stress–strain response, and photonic bandgap [159], and benefiting applications in catalyst [160], photonics [149], and metamaterials [161]. One of the promising strategies to realize such open lattice in an energy efficient way is by hierarchical assembly of patchy building blocks, which has also been exploited in polyhedral DNA scaffolds [162], collagen fibrils [163], and microscale particles [164]. The hierarchical assembly often requires two or more types of interparticle interactions that can be orthogonally triggered at different stages of assembly, which can be acquired by introducing patchiness. Both experimental and simulation efforts have been made to find the rules for achieving such high-level sophistication in superlattice. Muller and co-workers reported two types of polymeric patchy nanoparticles co-assemble into triangular clusters, which then further assemble into micron-scale compartmentalized chain or network structures, as dictated by the balance of interactions between attractive and repulsive patches [18, 51]. Chen and co-workers reported three types of hierarchically assembled structures: two different types of quasicrystals and body-centered-cubic supracrystals from patchy CdSe tetrahedral quantum dots, which are undergoing primary assembly into clusters attributed to specific facet-to-facet interaction between the patched surfaces [165, 166]. At an extremely small scale, the hierarchical construction of molecular films with honeycomb structure from fan-shaped molecular building blocks is achieved by Hou et al. (**Figure 10a–c**) [167]. The building units can be regarded as patchy as they are composed of polyoxometalate at the corner and polyhedral oligomeric silsesquioxanes on the one hand. On the other hand, using Monte Carlo simulations, rhombic platelets with two patches are triggered to assemble into several types of uniform clusters and into various lattices [168]. The structures were systematically explored as the simulation samples of the parameter space of attraction strength and patch position. Morphew et al. [169] used cluster-move Monte Carlo algorithm to understand how to encode hierarchy in interaction strength of triblock patchy particles. They further showed that hierarchy is essential for achieving final assembly of cubic diamond crystals via tetrahedral clusters (**Figure 10d**), as also consistent to the earlier experimental observation by Chen et al. (**Figure 10f**) [170]. When the patch width is expanded, the increasing patch-to-patch interaction range leads to the formation of body-centered cubic structure via octahedral clusters (**Figure 10e**), suggesting the generalizability of the design rules. The cooperative and iterative feedbacks shown in between this joint experiment and theory set a valuable example of how patchy nanoparticles assembly can effectively advance to introduce the novel hierarchical nanostructures.

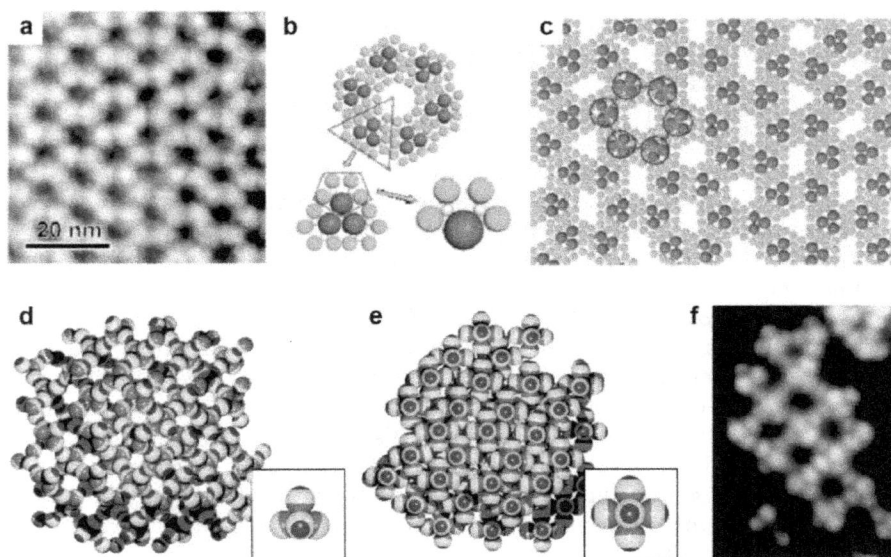

Figure 10.
(a) High-angle annular dark-field scanning transmission electron microscopy (HAADF-STEM) images showing honeycomb lattice composed of polyoxometalate (POM) cluster and four polyhedral oligomeric silsesquioxane (POSS) clusters (POM-4POSS). (b) Hierarchical structure of a honeycomb cell. Truncated triangle and the pentagon in red highlight the structural units at multilevels. (c) Snapshot of coarse-grained model simulation of honeycomb superstructure self-assembled from POM-4POSS. Adapted from Ref. [167]. Copyright 2018 American Chemical Society. (d and e) Two-level structural hierarchy in assembly structures from triblock patchy particle with varied patch size. Structural motifs (insets) and snapshot of typical crystal configuration assembled via (d) tetrahedral and (e) octahedral clusters. Adapted from Ref. [169]. Copyright 2018 American Chemical Society. (f) Fluorescence microscopy images of illustrative network structure assembled from triblock patchy particles. Adapted from Ref. [170]. Copyright 2012 American Chemical Society.

4. Conclusion and outlook

Diverse synthetic routes lead to different patchy nanoparticles with various patch size, number, morphology, and composition. To mention a few representative examples in terms of the composition, inorganic patchy particles can be obtained by controlled self-assembly of nanocrystals, the phase separation of polymer micelles leads to polymeric patchy particles, and the surface modification of inorganic nanoparticles can provide hybrid patchy nanoparticles with inorganic core and organic patches. In terms of self-assembly of patchy nanoparticles, different driving forces, including vdW, electrostatic, chemical and entropic interactions, have been explored. However, at current stage, most of the self-assembled structures are still limited to 0D cluster [30, 104] and 1D chain structures [15, 100, 101, 103]. Large-scale 2D planar structure [18] or 3D lattices [148] have only been realized in a few examples with limited domain size, causing a delay in realizing potential applications in medicine, electronic devices, and photonic crystals. Dispersity and impurity of the patchy nanoparticles are a few of main causes of this challenge. Thus, we expect patchy nanoparticles with polymeric patches could help, as the squish "soft" patches can easily adopt in the large-scale assembly structure and tolerate the differences in the shape and size of individual patchy particles.

On the other hand, direct imaging of patchy nanoparticle self-assembly in liquid can help advancing large scale assembly with low defects and long-range order. First of all, direct observation of assembly process provides quantitative "picture" of interparticle interaction governing dynamic assembly process. For example, from velocity and diffusivity of tracked nanoparticles forming chains and clusters, the magnitude and the range of forces can be experimentally investigated [171]. Moreover, as it also has been extensively demonstrated in micron-scale patchy particles under OM, real-time imaging provides a deep understanding in kinetic pathways of self-assembly process. For example, transient assembly structures before particles rearrange into final structures elucidates favored structures before equilibrium and the time scale of such events, providing the full assembly kinetic coordinates [172]. Thus the understanding of the kinetic coordinates in assembly can provide insights into generating highly ordered and desired assembly structures, by facilitating the specific assembly routes, among all other possible pathways upon differing assembly kinetics. Moreover, interference in transient assembly stages before the rearrangements, or restriction in assembly conditions can even "freeze" particles as they are trapped in nonequilibrium metastable structures, further opening future directions in achieving "kinetically-trapped" assembly structures by design [173]. Although OM has been routinely used to observe dynamics of micron-scale particles, it is not the case for the nanoparticles because of the diffraction limit of visible light disable single particle-level resolution at nanoscale. In order to circumvent such obstacle in terms of the imaging tool, recent progresses in liquid-phase transmission electron microscopy (LPTEM) have been gaining increasing attention, as it provides self-assembly pathways of nanoparticles in real time and space [88, 89, 174]. In order to push the limit of current status of LPTEM to fully resolve nanoparticle interactions within the time scale of assembly events, K3 cameras with a frame rate up to 1500 frames per second and machine learning algorithms to automatically extract out meaningful physical information from overwhelming amount of data have also been developed [175]. We see that with a proper handling of liquid confinement effect in chamber and the electron beam effect (e.g., the low image contrast and beam sensitivity of the organic patches), liquid-phase TEM technique can serve as a powerful tool to realize and experimentally guide patchy nanoparticle self-assembly for novel functional structures.

Acknowledgements

This work was financially supported by the National Science Foundation under Grant No. 1752517.

Author details

Ahyoung Kim[1], Lehan Yao[1], Falon Kalutantirige[2], Shan Zhou[1] and Qian Chen[1,2,3,4]*

1 Department of Materials Science and Engineering, University of Illinois at Urbana-Champaign, Urbana, Illinois, USA

2 Department of Chemistry, University of Illinois at Urbana-Champaign, Urbana, Illinois, USA

3 Beckman Institute for Advanced Science and Technology, University of Illinois at Urbana-Champaign, Urbana, Illinois, USA

4 Materials Research Laboratory, University of Illinois at Urbana-Champaign, Urbana, Illinois, USA

*Address all correspondence to: qchen20@illinois.edu

IntechOpen

References

[1] Yake AM, Snyder CE, Velegol D. Site-specific functionalization on individual colloids: Size control, stability, and multilayers. Langmuir. 2007;**23**(17):9069-9075

[2] Roh K-H, Martin DC, Lahann J. Triphasic nanocolloids. Journal of the American Chemical Society. 2006;**128**(21):6796-6797

[3] Roh K-H, Martin DC, Lahann J. Biphasic Janus particles with nanoscale anisotropy. Nature Materials. 2005;**4**(10):759-763

[4] Perro A, Reculusa S, Ravaine S, Bourgeat-Lami E, Duguet E. Design and synthesis of janus micro- and nanoparticles. Journal of Materials Chemistry. 2005;**15**(35-36):3745-3760

[5] Zlotnick A, Stray SJ. How does your virus grow? Understanding and interfering with virus assembly. Trends in Biotechnology. 2003;**21**(12):536-542

[6] Chakrabarti P. Dissecting protein–protein recognition sites. 2002;**343**(August 2001):334-343

[7] Glotzer SC, Solomon MJ. Anisotropy of building blocks and their assembly into complex structures. Nature Materials. 2007;**6**(8):557-562

[8] Zhang Z, Glotzer SC. Self-assembly of patchy particles. Nano Letters. 2004;**4**(8):1407-1413

[9] Pawar AB, Kretzschmar I. Fabrication, assembly, and application of patchy particles. Macromolecular Rapid Communications. 2010;**31**(2):150-168

[10] Sciortino F, Tartaglia P. Glassy colloidal systems. Advances in Physics. 2005;**54**(6-7):471-524

[11] Kim J-W, Larsen RJ, Weitz DA. Uniform nonspherical colloidal

particles with tunable shapes. Advanced Materials. 2007;**19**(15):2005-2009

[12] Klupp Taylor RN, Bao H, Tian C, Vasylyev S, Peukert W. Facile route to morphologically tailored silver patches on colloidal particles. Langmuir. 2010;**26**(16):13564-13571

[13] Kim A, Zhou S, Yao L, Ni S, Luo B, Sing CE, et al. Tip-patched nanoprisms from formation of Ligand Islands. Journal of the American Chemical Society. 2019;**141**(30):11796-11800

[14] Zhu S, Li Z-W, Zhao H. Patchy micelles based on coassembly of block copolymer chains and block copolymer brushes on silica particles. Langmuir. 2015;**31**(14):4129-4136

[15] Nie Z, Fava D, Kumacheva E, Zou S, Walker GC, Rubinstein M. Self-assembly of metal–polymer analogues of amphiphilic triblock copolymers. Nature Materials. 2007;**6**(8):609-614

[16] Tan SF, Anand U, Mirsaidov U. Interactions and attachment pathways between functionalized gold nanorods. ACS Nano. 2017;**11**(2):1633-1640

[17] Wang T, Zhuang J, Lynch J, Chen O, Wang Z, Wang X, et al. Self-assembled colloidal superparticles from nanorods. Science. 2012;**338**(6105):358-363

[18] Gröschel AH, Walther A, Löbling TI, Schacher FH, Schmalz H, Müller AHE. Guided hierarchical co-assembly of soft patchy nanoparticles. Nature. 2013;**503**(7475):247-251

[19] Coluzza I, Van Oostrum PDJ, Capone B, Reimhult E, Dellago C. Sequence controlled self-knotting colloidal patchy polymers. Physical Review Letters. 2013;**110**(7):1-5

[20] Gangwal S, Pawar A, Velev OD. Programmed assembly of

metallodielectric patchy particles in external AC electric fields. Soft Review. 2010;**6**:1413-1418

[21] Feng L, Dreyfus R, Sha R, Seeman NC, Chaikin PM. DNA patchy particles. Advanced Materials. 2013;**25**(20):2779-2783

[22] Grzelczak M, Vermant J, Furst EM, Liz-Marzán LM. Directed self-assembly of nanoparticles. ACS Nano. 2010;**4**(7):3591-3605

[23] Lee YH, Lay CL, Shi W, Lee HK, Yang Y, Li S, et al. Creating two self-assembly micro-environments to achieve supercrystals with dual structures using polyhedral nanoparticles. Nature Communications. 2018;**9**(1):1-8

[24] Damasceno PF, Engel M, Glotzer SC. Predictive self-assembly of polyhedra into complex structures. Science. 2012;**337**(6093):453-457

[25] Evers WH, Goris B, Bals S, Casavola M, de Graaf J, van Roij R, et al. Low-dimensional semiconductor superlattices formed by geometric control over nanocrystal attachment. Nano Letters. 2013;**13**(6):2317-2323

[26] Young KL, Personick ML, Engel M, Damasceno PF, Barnaby SN, Bleher R, et al. A directional entropic force approach to assemble anisotropic nanoparticles into superlattices. Angewandte Chemie, International Edition. 2013;**52**(52):13980-13984

[27] Cersonsky RK, van Anders G, Dodd PM, Glotzer SC. Relevance of packing to colloidal self-assembly. PNAS. 2018;**115**(7):1439-1444

[28] Kang C, Honciuc A. Influence of geometries on the assembly of snowman-shaped Janus nanoparticles. ACS Nano. 2018;**12**(4):3741-3750

[29] Hu H, Ji F, Xu Y, Yu J, Liu Q, Chen L, et al. Reversible and precise self-assembly of Janus metal-organosilica nanoparticles through a linker-free approach. ACS Nano. 2016;**10**(8):7323-7330

[30] Xu X, Rosi NL, Wang Y, Huo F, Mirkin CA. Asymmetric functionalization of gold nanoparticles with oligonucleotides. Journal of the American Chemical Society. 2006;**128**(29):9286-9287

[31] Liu W, Halverson J, Tian Y, Tkachenko AV, Gang O. Self-organized architectures from assorted DNA-framed nanoparticles. Nature Chemistry. 2016;**8**(9):867-873

[32] Chen Q, Diesel E, Whitmer JK, Bae SC, Luijten E, Granick S. Triblock colloids for directed self-assembly. Journal of the American Chemical Society. 2011;**133**(20):7725-7727

[33] Doppelbauer G, Noya EG, Bianchi E, Kahl G. Self-assembly scenarios of patchy colloidal particles. Soft Matter. 2012;**8**(30):7768-7772

[34] Marín-Aguilar S, Wensink HH, Foffi G, Smallenburg F. Slowing down supercooled liquids by manipulating their local structure. Soft Matter. 2019;**15**(48):9886-9893

[35] Słyk E, Rżysko W, Bryk P. Two-dimensional binary mixtures of patchy particles and spherical colloids. Soft Matter. 2016;**12**(47):9538-9548

[36] Chen Q, Bae SC, Granick S. Directed self-assembly of a colloidal Kagome lattice. Nature. 2011;**469**(7330):381-384

[37] Rocklin DZ, Mao X. Self-assembly of three-dimensional open structures using patchy colloidal particles. Soft Matter. 2014;**10**(38):7569-7576

[38] Wilber AW, Doye JPK, Louis AA, Noya EG, Miller MA, Wong P. Reversible self-assembly of patchy

particles into monodisperse icosahedral clusters. The Journal of Chemical Physics. 2007;**127**(8):085106

[39] Mao X, Chen Q, Granick S. Entropy favours open colloidal lattices. Nature Materials. 2013;**12**(3):217-222

[40] Stolarczyk JK, Deak A, Brougham DF. Nanoparticle clusters: Assembly and control over internal order, current capabilities, and future potential. Advanced Materials. 2016;**28**(27):5400-5424

[41] Manoharan VN, Elsesser MT, Pine DJ. Dense packing and symmetry in small clusters of microspheres. Science. 2003;**301**(5632):483-487

[42] Cho Y-S, Yi G-R, Lim J-M, Kim S-H, Manoharan VN, Pine DJ, et al. Self-organization of bidisperse colloids in water droplets. Journal of the American Chemical Society. 2005;**127**(45):15968-15975

[43] Wagner CS, Lu Y, Wittemann A. Preparation of submicrometer-sized clusters from polymer spheres using ultrasonication. Langmuir. 2008;**24**(21):12126-12128

[44] Wang Y, Chen G, Yang M, Silber G, Xing S, Tan LH, et al. A systems approach towards the stoichiometry-controlled hetero-assembly of nanoparticles. Nature Communications. 2010;**1**(1):87

[45] Xiong S, Xi B, Zhang K, Chen Y, Jiang J, Hu J, et al. Ag nanoprisms with Ag 2 S attachment. Scientific Reports. 2013;**3**(1):1-9

[46] Huang Z, Zhao Z-J, Zhang Q, Han L, Jiang X, Li C, et al. A welding phenomenon of dissimilar nanoparticles in dispersion. Nature Communications. 2019;**10**(1):1-8

[47] Mastroianni AJ, Claridge SA, Alivisatos AP. Pyramidal and chiral

groupings of gold nanocrystals assembled using DNA scaffolds. Journal of the American Chemical Society. 2009;**131**(24):8455-8459

[48] Xue L, Zhang J, Han Y. Phase separation induced ordered patterns in thin polymer blend films. Progress in Polymer Science. 2012;**37**(4):564-594

[49] Kim DH, Kim WT, Park ES, Mattern N, Eckert J. Phase separation in metallic glasses. Progress in Materials Science. 2013;**58**(8):1103-1172

[50] Gröschel AH, Walther A, Löbling TI, Schmelz J, Hanisch A, Schmalz H, et al. Facile, solution-based synthesis of soft, nanoscale Janus particles with tunable Janus balance. Journal of the American Chemical Society. 2012;**134**(33):13850-13860

[51] Gröschel AH, Schacher FH, Schmalz H, Borisov OV, Zhulina EB, Walther A, et al. Precise hierarchical self-assembly of multicompartment micelles. Nature Communications. 2012;**3**(1):1-10

[52] Deng R, Liang F, Qu X, Wang Q, Zhu J, Yang Z. Diblock copolymer based Janus nanoparticles. Macromolecules. 2015;**48**(3):750-755

[53] Deng R, Liu S, Liang F, Wang K, Zhu J, Yang Z. Polymeric Janus particles with hierarchical structures. Macromolecules. 2014;**47**(11):3701-3707

[54] Chen P-C, Liu X, Hedrick JL, Xie Z, Wang S, Lin Q-Y, et al. Polyelemental nanoparticle libraries. Science. 2016;**352**(6293):1565-1569

[55] Singh C, Ghorai PK, Horsch MA, Jackson AM, Larson RG, Stellacci F, et al. Entropy-mediated patterning of surfactant-coated nanoparticles and surfaces. Physical Review Letters. 2007;**99**(22):226106

[56] Chen T, Yang M, Wang X, Tan LH, Chen H. Controlled assembly

of eccentrically encapsulated gold nanoparticles. Journal of the American Chemical Society. 2008;**130**(36):11858-11859

[57] Tan LH, Xing H, Chen H, Lu Y. Facile and efficient preparation of anisotropic DNA-functionalized gold nanoparticles and their regioselective assembly. Journal of the American Chemical Society. 2013;**135**(47):17675-17678

[58] Choueiri RM, Galati E, Thérien-Aubin H, Klinkova A, Larin EM, Querejeta-Fernández A, et al. Surface patterning of nanoparticles with polymer patches. Nature. 2016;**538**(7623):79-83

[59] Galati E, Tebbe M, Querejeta-Fernández A, Xin HL, Gang O, Zhulina EB, et al. Shape-specific patterning of polymer-functionalized nanoparticles. ACS Nano. 2017;**11**(5):4995-5002

[60] Tao H, Chen L, Galati E, Manion JG, Seferos DS, Zhulina EB, et al. Helicoidal patterning of gold nanorods by phase separation in mixed polymer brushes. Langmuir. 2019;**35**(48):15872-15879

[61] Zeng J, Zhu C, Tao J, Jin M, Zhang H, Li ZY, et al. Controlling the nucleation and growth of silver on palladium nanocubes by manipulating the reaction kinetics. Angewandte Chemie, International Edition. 2012;**51**(10):2354-2358

[62] Mokari T, Rothenberg E, Popov I, Costi R, Banin U. Selective growth of metal tips onto semiconductor quantum rods and tetrapods. Science. 2004;**304**(5678):1787-1790

[63] Buonsanti R, Grillo V, Carlino E, Giannini C, et al. Seeded growth of asymmetric binary nanocrystals made of a semiconductor TiO_2 rodlike section and a magnetic

γ-Fe_2O_3 spherical domain. Journal of the American Chemical Society. 2006;**128**(51):16953-16970

[64] Robinson RD, Sadtler B, Demchenko DO, Erdonmez CK, Wang L-W, Alivisatos AP. Spontaneous superlattice formation in nanorods through partial cation exchange. Science. 2007;**317**(5836):355-358

[65] Kwon SG, Krylova G, Phillips PJ, Klie RF, Chattopadhyay S, Shibata T, et al. Heterogeneous nucleation and shape transformation of multicomponent metallic nanostructures. Nature Materials. 2015;**14**(2):215-223

[66] Mazumder V, Chi M, More KL, Sun S. Synthesis and characterization of multimetallic Pd/Au and Pd/Au/FePt core/shell nanoparticles. Angewandte Chemie International Edition. 2010;**49**(49):9368-9372

[67] Mokari T, Sztrum CG, Salant A, Rabani E, Banin U. Formation of asymmetric one-sided metal-tipped semiconductor nanocrystal dots and rods. Nature Materials. 2005;**4**(11):855-863

[68] Halpert JE, Porter VJ, Zimmer JP, Bawendi MG. Synthesis of CdSe/CdTe nanobarbells. Journal of the American Chemical Society. 2006;**128**(39):12590-12591

[69] Krylova G, Giovanetti LJ, Requejo FG, Dimitrijevic NM, Prakapenka A, Shevchenko EV. Study of nucleation and growth mechanism of the metallic nanodumbbells. Journal of the American Chemical Society. 2012;**134**(9):4384-4392

[70] Peng S, Lei C, Ren Y, Cook RE, Sun Y. Plasmonic/magnetic bifunctional nanoparticles. Angewandte Chemie, International Edition. 2011;**50**(14):3158-3163

[71] Schick I, Lorenz S, Gehrig D, Schilmann AM, Bauer H, Panthöfer M, et al. Multifunctional two-photon active silica-coated Au@MnO Janus particles for selective dual functionalization and imaging. Journal of the American Chemical Society. 2014;**136**(6):2473-2483

[72] Zeng J, Huang J, Liu C, Wu CH, Lin Y, Wang X, et al. Gold-based hybrid nanocrystals through heterogeneous nucleation and growth. Advanced Materials. 2010;**22**(17):1936-1940

[73] Wang F, Cheng S, Bao Z, Wang J. Anisotropic overgrowth of metal heterostructures induced by a site-selective silica coating. Angewandte Chemie—International Edition. 2013;**52**(39):10344-10348

[74] Wu B, Liu D, Mubeen S, Chuong TT, Moskovits M, Stucky GD. Anisotropic growth of TiO_2 onto gold nanorods for plasmon-enhanced hydrogen production from water reduction. Journal of the American Chemical Society. 2016;**138**(4):1114-1117

[75] Szekrényes DP, Pothorszky S, Zámbó D, Osváth Z, Deák A. Investigation of patchiness on tip-selectively surface-modified gold nanorods. The Journal of Physical Chemistry C. 2018;**122**(3):1706-1710

[76] Nepal D, Onses MS, Park K, Jespersen M, Thode CJ, Nealey PF, et al. Control over position, orientation, and spacing of arrays of gold nanorods using chemically nanopatterned surfaces and tailored particle–particle–surface interactions. ACS Nano. 2012;**6**(6):5693-5701

[77] He W, Frueh J, Wu Z, He Q. Leucocyte membrane-coated Janus microcapsules for enhanced photothermal cancer treatment. Langmuir. 2016;**32**(15):3637-3644

[78] Pan Y, Gao J, Zhang B, Zhang X, Xu B. Colloidosome-based synthesis of a multifunctional nanostructure of silver and hollow iron oxide nanoparticles. Langmuir. 2010;**26**(6):4184-4187

[79] Gu H, Yang Z, Gao J, Chang CK, Xu B. Heterodimers of nanoparticles: Formation at a liquid–liquid interface and particle-specific surface modification by functional molecules. Journal of the American Chemical Society. 2005;**127**(1):34-35

[80] He J, Perez MT, Zhang P, Liu Y, Babu T, Gong J, et al. A general approach to synthesize asymmetric hybrid nanoparticles by interfacial reactions. Journal of the American Chemical Society. 2012;**134**(8):3639-3642

[81] Ayala A, Carbonell C, Imaz I, Maspoch D. Introducing asymmetric functionality into MOFs via the generation of metallic Janus MOF particles. Chemical Communications. 2016;**52**(29):5096-5099

[82] Ma X, Sanchez S. A bio-catalytically driven Janus mesoporous silica cluster motor with magnetic guidance. Chemical Communications. 2015;**51**(25):5467-5470

[83] Yin Y, Zhou S, You B, Wu L. Facile fabrication and self-assembly of polystyrene–silica asymmetric colloid spheres. Journal of Polymer Science Part A: Polymer Chemistry. 2011;**49**(15):3272-3279

[84] Gong J, Zu X, Li Y, Mu W, Deng Y. Janus particles with tunable coverage of zinc oxide nanowires. Journal of Materials Chemistry. 2011;**21**(7):2067-2069

[85] Huang Z, Liu Y, Zhang Q, Chang X, Li A, Deng L, et al. Collapsed polymer-directed synthesis of multicomponent coaxial-like nanostructures. Nature Communications. 2016;**7**(1):12147

[86] Zhang, Keys AS, Chen T, Glotzer SC. Self-assembly of patchy

particles into diamond structures through molecular mimicry. Langmuir. 2005;**21**(25):11547-11551

[87] Zhang J, Luijten E, Granick S. Toward design rules of directional Janus colloidal assembly. Annual Review of Physical Chemistry. 2015;**66**(1):581-600

[88] Kim J, Ou Z, Jones MR, Song X, Chen Q. Imaging the polymerization of multivalent nanoparticles in solution. Nature Communications. 2017;**8**(1):1-10

[89] Ou Z, Wang Z, Luo B, Luijten E, Chen Q. Kinetic pathways of crystallization at the nanoscale. Nature Materials. 2019:1-6

[90] Batista CAS, Larson RG, Kotov NA. Nonadditivity of nanoparticle interactions. Science. 2015;**350**(6257)

[91] Luo B, Smith JW, Wu Z, Kim J, Ou Z, Chen Q. Polymerization-like co-assembly of silver nanoplates and patchy spheres. ACS Nano. 2017;**11**(8):7626-7633

[92] Verwey EJW. Theory of the stability of lyophobic colloids. The Journal of Physical Chemistry. 1947;**51**(3):631-636

[93] Zhu H, Fan Z, Yu L, Wilson MA, Nagaoka Y, Eggert D, et al. Controlling nanoparticle orientations in the self-assembly of patchy quantum dot-gold heterostructural nanocrystals. Journal of the American Chemical Society. 2019;**141**(14):6013-6021

[94] Li W, Palis H, Mérindol R, Majimel J, Ravaine S, Duguet E. Colloidal molecules and patchy particles: complementary concepts, synthesis and self-assembly. Chemical Society Reviews. 2020;**49**(6):1955-1976

[95] van Oss CJ. Chapter 1: General and historical introduction in interface science and technology. In: van Oss CJ, editor. The Properties of Water and their Role in Colloidal and Biological Systems.

Vol. 16. Amsterdam, The Netherlands: Elsevier; 2008. pp. 1-9

[96] Nie Z, Fava D, Rubinstein M, Kumacheva E. "Supramolecular" assembly of gold nanorods end-terminated with polymer "Pom-Poms": Effect of Pom-Pom structure on the association modes. Journal of the American Chemical Society. 2008;**130**(11):3683-3689

[97] Liu K, Nie Z, Zhao N, Li W, Rubinstein M, Kumacheva E. Step-growth polymerization of inorganic nanoparticles. Science. 2010;**329**(5988):197-200

[98] Liu K, Lukach A, Sugikawa K, Chung S, Vickery J, Therien-Aubin H, et al. Copolymerization of metal nanoparticles: A route to colloidal plasmonic copolymers. Angewandte Chemie, International Edition. 2014;**53**(10):2648-2653

[99] Klinkova A, Thérien-Aubin H, Choueiri RM, Rubinstein M, Kumacheva E. Colloidal analogs of molecular chain stoppers. PNAS. 2013;**110**(47):18775-18779

[100] Kim J-H, Jong Kwon W, Sohn B-H. Supracolloidal polymer chains of Diblock copolymer micelles. ChemComm. 2015;**51**(16):3324-3327

[101] Cui H, Chen Z, Zhong S, Wooley KL, Pochan DJ. Block copolymer assembly via kinetic control. Science. 2007;**317**(5838):647-650

[102] Yi G-R, Pine DJ, Sacanna S. Recent progress on patchy colloids and their self-assembly. Journal of Physics: Condensed Matter. 2013;**25**(19):193101

[103] Caswell KK, Wilson JN, Bunz UHF, Murphy CJ. Preferential end-to-end assembly of gold nanorods by biotin–streptavidin connectors. Journal of the American Chemical Society. 2003;**125**(46):13914-13915

[104] Rouet P-E, Chomette C, Adumeau L, Duguet E, Ravaine S. Colloidal chemistry with patchy silica nanoparticles. Beilstein Journal of Nanotechnology. 2018;**9**(1):2989-2998

[105] Mirkin CA, Letsinger RL, Mucic RC, Storhoff JJ. A DNA-based method for rationally assembling nanoparticles into macroscopic materials. Nature. 1996;**382**(6592):607-609

[106] Alivisatos AP, Johnsson KP, Peng X, Wilson TE, Loweth CJ, Bruchez MP, et al. Organization of "nanocrystal molecules" using DNA. Nature. 1996;**382**(6592):609-611

[107] Liu W, Tagawa M, Xin HL, Wang T, Emamy H, Li H, et al. Diamond family of nanoparticle superlattices. Science. 2016;**351**(6273):582-586

[108] Chen G, Gibson KJ, Liu D, Rees HC, Lee J-H, Xia W, et al. Regioselective surface encoding of nanoparticles for programmable self-assembly. Nature Materials. 2019;**18**(2):169-174

[109] Asakura S, Oosawa F. On interaction between two bodies immersed in a solution of macromolecules. The Journal of Chemical Physics. 1954;**22**(7):1255-1256

[110] Götzelmann B, Evans R, Dietrich S. Depletion forces in fluids. Physical Review E. 1998;**57**(6):6785-6800

[111] Kim S-H, Hollingsworth AD, Sacanna S, Chang S-J, Lee G, Pine DJ, et al. Synthesis and assembly of colloidal particles with sticky dimples. Journal of the American Chemical Society. 2012;**134**(39):16115-16118

[112] Wang Y, Wang Y, Zheng X, Yi G-R, Sacanna S, Pine DJ, et al. Three-dimensional lock and key colloids. Journal of the American Chemical Society. 2014;**136**(19):6866-6869

[113] Sacanna S, Irvine WTM, Chaikin PM, Pine DJ. Lock and key colloids. Nature. 2010;**464**(7288):575-578

[114] Kim J, Song X, Ji F, Luo B, Ice NF, Liu Q, et al. Polymorphic assembly from beveled gold triangular nanoprisms. Nano Letters. 2017;**17**(5):3270-3275

[115] van Anders G, Klotsa D, Ahmed NK, Engel M, Glotzer SC. Understanding shape entropy through local dense packing. PNAS. 2014;**111**(45):E4812-E4821

[116] Harper ES, van Anders G, Glotzer SC. The entropic bond in colloidal crystals. PNAS. 2019;**116**(34):16703-16710

[117] Geng Y, van Anders G, Dodd PM, Dshemuchadse J, Glotzer SC. Engineering entropy for the inverse design of colloidal crystals from hard shapes. Science Advances. 2019;**5**(7):eaaw0514

[118] van Anders G, Ahmed NK, Smith R, Engel M, Glotzer SC. Entropically patchy particles: Engineering valence through shape entropy. ACS Nano. 2014;**8**(1):931-940

[119] Petukhov AV, Tuinier R, Vroege GJ. Entropic patchiness: Effects of colloid shape and depletion. Current Opinion in Colloid & Interface Science. 2017;**30**:54-61

[120] Damasceno PF, Engel M, Glotzer SC. Crystalline assemblies and densest packings of a family of truncated tetrahedra and the role of directional entropic forces. ACS Nano. 2012;**6**(1):609-614

[121] Ye X, Collins JE, Kang Y, Chen J, Chen DTN, Yodh AG, et al. Morphologically controlled synthesis of colloidal upconversion nanophosphors and their shape-directed self-assembly. PNAS. 2010;**107**(52):22430-22435

[122] Ye X, Chen J, Engel M, Millan JA, Li W, Qi L, et al. Competition of shape and interaction patchiness for self-assembling nanoplates. Nature Chemistry. 2013;**5**(6):466-473

[123] Talapin DV, Shevchenko EV, Bodnarchuk MI, Ye X, Chen J, Murray CB. Quasicrystalline order in self-assembled binary nanoparticle superlattices. Nature. 2009;**461**(7266):964-967

[124] Ye X, Millan JA, Engel M, Chen J, Diroll BT, Glotzer SC, et al. Shape alloys of nanorods and nanospheres from self-assembly. Nano Letters. 2013;**13**(10):4980-4988

[125] Yan J, Bae SC, Granick S. Colloidal superstructures programmed into magnetic Janus particles. Advanced Materials. 2015;**27**(5):874-879

[126] Lumsdon SO, Kaler EW, Velev OD. Two-dimensional crystallization of microspheres by a coplanar AC electric field. Langmuir. 2004;**20**(6):2108-2116

[127] Song P, Wang Y, Wang Y, Hollingsworth AD, Weck M, Pine DJ, et al. Patchy particle packing under electric fields. Journal of the American Chemical Society. 2015;**137**(8):3069-3075

[128] Ma F, Wang S, Wu DT, Wu N. Electric-field–induced assembly and propulsion of chiral colloidal clusters. PNAS. 2015;**112**(20):6307-6312

[129] Zhang J, Yan J, Granick S. Directed self-assembly pathways of active colloidal clusters. Angewandte Chemie, International Edition. 2016;**55**(17):5166-5169

[130] Yan J, Han M, Zhang J, Xu C, Luijten E, Granick S. Reconfiguring active particles by electrostatic imbalance. Nature Materials. 2016;**15**(10):1095-1099

[131] Lee JG, Brooks AM, Shelton WA, Bishop KJM, Bharti B. Directed propulsion of spherical particles along three dimensional helical trajectories. Nature Communications. 2019;**10**(1):1-8

[132] Yan J, Chaudhary K, Chul Bae S, Lewis JA, Granick S. Colloidal ribbons and rings from Janus magnetic rods. Nature Communications. 2013;**4**(1):1-9

[133] Smoukov SK, Gangwal S, Marquez M, Velev OD. Reconfigurable responsive structures assembled from magnetic Janus particles. Soft Matter. 2009;**5**(6):1285-1292

[134] Ren B, Ruditskiy A, Song JH(K), Kretzschmar I. Assembly behavior of Iron oxide-capped Janus particles in a magnetic field. Langmuir. 2012;**28**(2):1149-1156

[135] Sacanna S, Rossi L, Pine DJ. Magnetic click colloidal assembly. Journal of the American Chemical Society. 2012;**134**(14):6112-6115

[136] Yan J, Bloom M, Bae SC, Luijten E, Granick S. Linking synchronization to self-assembly using magnetic Janus colloids. Nature. 2012;**491**(7425):578-581

[137] Singh G, Chan H, Baskin A, Gelman E, Repnin N, Král P, et al. Self-assembly of magnetite nanocubes into helical superstructures. Science. 2014;**345**(6201):1149-1153

[138] Fu Z, Xiao Y, Feoktystov A, Pipich V, et al. Field-induced self-assembly of iron oxide nanoparticles investigated using small-angle neutron scattering. Nanoscale. 2016;**8**(43):18541-18550

[139] Luo B, Smith JW, Ou Z, Chen Q. Quantifying the self-assembly behavior of anisotropic nanoparticles using liquid-phase transmission electron microscopy. Accounts of Chemical Research. 2017;**50**(5):1125-1133

[140] Kim J, Jones MR, Ou Z, Chen Q. In situ electron microscopy imaging and quantitative structural modulation of nanoparticle superlattices. ACS Nano. 2016;**10**(11):9801-9808

[141] Zheng H. Using molecular tweezers to move and image nanoparticles. Nanoscale. 2013;**5**(10):4070-4078

[142] Rovigatti L, Russo J, Romano F. How to simulate patchy particles. European Physical Journal E: Soft Matter and Biological Physics. 2018;**41**(5):59

[143] Ranguelov B, Nanev C. 2D Monte Carlo simulation of patchy particles association and protein crystal polymorph selection. Crystals. 2019;**9**(10):508

[144] Carrillo-Tripp M, Shepherd CM, et al. VIPERdb2: An enhanced and web API enabled relational database for structural virology. Nucleic Acids Research. 2009;**37**(suppl_1):D436-D442

[145] Perlmutter JD, Hagan MF. Mechanisms of virus assembly. Annual Review of Physical Chemistry. 2015;**66**(1):217-239

[146] Smallenburg F, Sciortino F. Liquids more stable than crystals in particles with limited valence and flexible bonds. Nature Physics. 2013;**9**(9):554-558

[147] Lu F, Vo T, Zhang Y, Frenkel A, Yager KG, Kumar S, et al. Unusual packing of soft-shelled nanocubes. Science Advances. 2019;**5**(5):eaaw2399

[148] Walker DA, Leitsch EK, Nap RJ, Szleifer I, Grzybowski BA. Geometric curvature controls the chemical patchiness and self-assembly of nanoparticles. Nature Nanotechnology. 2013;**8**(9):676-681

[149] Romano F, Sciortino F. Patterning symmetry in the rational design of colloidal crystals. Nature Communications. 2012;**3**(1):1-6

[150] Guo R, Mao J, Xie X-M, Yan L-T. Predictive supracolloidal helices from patchy particles. Scientific Reports. 2014;**4**(1):1-7

[151] Gemeinhardt A, Martinsons M, Schmiedeberg M. Growth of two-dimensional dodecagonal colloidal quasicrystals: Particles with isotropic pair interactions with two length scales vs. patchy colloids with preferred binding angles. European Physical Journal E: Soft Matter and Biological Physics. 2018;**41**(10):126

[152] Mahynski NA, Rovigatti L, Likos CN, Panagiotopoulos AZ. Bottom-up colloidal crystal assembly with a twist. ACS Nano. 2016;**10**(5):5459-5467

[153] Karner C, Dellago C, Bianchi E. Design of patchy rhombi: From close-packed tilings to open lattices. Nano Letters. 2019;**19**(11):7806-7815

[154] Rocklin DZ, Zhou S, Sun K, Mao X. Transformable topological mechanical metamaterials. Nature Communications. 2017;**8**(1):1-9

[155] Sun K, Souslov A, Mao X, Lubensky TC. Surface phonons, elastic response, and conformal invariance in twisted Kagome lattices. PNAS. 2012;**109**(31):12369-12374

[156] Ma Y, Ferguson AL. Inverse design of self-assembling colloidal crystals with omnidirectional photonic bandgaps. Soft Matter. 2019;**15**(43):8808-8826

[157] Chen D, Zhang G, Torquato S. Inverse design of colloidal crystals via optimized patchy interactions. The Journal of Physical Chemistry. B. 2018;**122**(35):8462-8468

[158] Whitelam S. Minimal positive Design for self-assembly of the archimedean tilings. Physical Review Letters. 2016;**117**(22):228003

[159] Jia Z, Liu F, Jiang X, Wang L. Engineering lattice metamaterials for extreme property, programmability, and multifunctionality. Journal of Applied Physics. 2020;**127**(15):150901

[160] Nai J, Wang S, Lou XW(D). Ordered colloidal clusters constructed by nanocrystals with valence for efficient CO2 photoreduction. Science Advances. 2019;**5**(12):eaax5095

[161] Kadic M, Milton GW, van Hecke M, Wegener M. 3D metamaterials. Nature Reviews Physics. 2019;**1**(3):198-210

[162] He Y, Ye T, Su M, Zhang C, Ribbe AE, Jiang W, et al. Hierarchical self-assembly of DNA into symmetric supramolecular polyhedra. Nature. 2008;**452**(7184):198-201

[163] Minary-Jolandan M, Yu M-F. Nanomechanical heterogeneity in the gap and overlap regions of type I collagen fibrils with implications for bone heterogeneity. Biomacromolecules. 2009;**10**(9):2565-2570

[164] Luo B, Kim A, Smith JW, Ou Z, Wu Z, Kim J, et al. Hierarchical self-assembly of 3D lattices from polydisperse anisometric colloids. Nature Communications. 2019;**10**(1):1-9

[165] Nagaoka Y, Tan R, Li R, Zhu H, Eggert D, Wu YA, et al. Superstructures generated from truncated tetrahedral quantum dots. Nature. 2018;**561**(7723):378-382

[166] Nagaoka Y, Zhu H, Eggert D, Chen O. Single-component quasi-crystalline nanocrystal superlattices through flexible polygon tiling rule. Science. 2018;**362**(6421):1396-1400

[167] Hou X-S, Zhu G-L, Ren L-J, Huang Z-H, Zhang R-B, Ungar G, et al. Mesoscale graphene-like honeycomb mono- and multilayers constructed via self-assembly of coclusters. Journal of the American Chemical Society. 2018;**140**(5):1805-1811

[168] Karner C, Dellago C, Bianchi E. Hierarchical self-assembly of patchy colloidal platelets. Soft Matter. 2020;**16**(11):2774-2785

[169] Morphew D, Shaw J, Avins C, Chakrabarti D. Programming hierarchical self-assembly of patchy particles into colloidal crystals via colloidal molecules. ACS Nano. 2018;**12**(3):2355-2364

[170] Chen Q, Bae SC, Granick S. Staged self-assembly of colloidal metastructures. Journal of the American Chemical Society. 2012;**134**(27):11080-11083

[171] Powers AS, Liao H-G, Raja SN, Bronstein ND, Alivisatos AP, Zheng H. Tracking nanoparticle diffusion and interaction during self-assembly in a liquid cell. Nano Letters. 2017;**17**(1):15-20

[172] Chen Q, Whitmer JK, Jiang S, Bae SC, Luijten E, Granick S. Supracolloidal reaction kinetics of Janus spheres. Science. 2011;**331**(6014):199-202

[173] Lunn DJ, Finnegan JR, Manners I. Self-assembly of "patchy" nanoparticles: A versatile approach to functional hierarchical materials. Chemical Science. 2015;**6**(7):3663-3673

[174] Liu C, Ou Z, Guo F, Luo B, Chen W, Qi L, et al. "Colloid–atom duality" in the assembly dynamics of concave gold nanoarrows. Journal of the American Chemical Society. 2020;**142**(27):11669-11673

[175] Yao L, Ou Z, Luo B, Xu C, Chen Q. Machine learning to reveal nanoparticle dynamics from liquid-phase TEM videos. ACS Central Science. 2020. DOI: 10.1021/acscentsci.0c00430

Chapter 5

Self-Assembled Copper Polypyridyl Supramolecular Metallopolymer Achieving Enhanced Anticancer Efficacy

Zushuang Xiong, Lanhai Lai and Tianfeng Chen

Abstract

Metallopolymers, a combination of organic polymers and metal center, contain metal atoms in repeating monomers can change its dynamic and thermodynamic properties through the directionality of coordination bonds and chemical tailoring of ligands. In the past decade, self-assembled functional supramolecular metallopolymers have aroused a surge of research interest, and have demonstrated application potential in cancer therapy. In this chapter, we have summarized the progress in the rational design of biological application of different metallopolymers. Especially, a copper polypyridyl complex was found be able to self-assemble into a supramolecular metallopolymer driven by the intermolecular interactions, which could enhance the uptake in cancer cells through endocytosis, thus effectively inhibit tumor growth in vivo without damage to the major organs. This study may provide a good example to use self-assembled metallopolymer to achieve enhanced anticancer efficacy.

Keywords: progress in self-assemble, metallopolymer, anticancer, copper (II) complex

1. Introduction

Metallopolymers, a combination of organic polymers and metal center, contain metal atoms in repeating monomers can regulate its dynamic stability and thermodynamic properties through the variation of coordination mode of central-metal ions and chemical tailoring of ligands [1]. With purpose to improve material properties, metallopolymers have attracted increasing interest for their potential to supply advanced functional materials for a wide range of applications [2]. Supramolecular polymer, originating from the integration of polymer and supramolecule, is becoming a rapidly developing research area in recent decades [3–7]. Supramolecular metallopolymers received increasing attention, partly motivated by their ready-to-form self-assembly [8–10] and diverse applications in electrochromic materials [11], luminescent [12–14], accelerated guest adsorption [15], interesting magnetic properties [16], and so on [9, 17]. Nitschke et al. synthesized a new type of metallopolymer, by exploring the gel self-assembly process and formation conditions, and explored the electrochemical properties, photoluminescence properties and thermochromism of the polymer [12]. Che and his co-workers synthesized cyclometalated Au^III complexes, which can be self-assembled to form supramolecular polymer, through the hydrogen bonding of the guanine-like amino group of the 4-DPT ligand and the π-π stacking interaction of 2,6-diphenylpyridine. The selective growth inhibition of supramolecular polymers on tumor cells and its possible mechanism were also investigated (**Figure 1**) [17].

Weak noncovalent interactions, such as hydrogen bonding, hydrophobic-hydrophobic, metal-metal and π-π interactions [18–24], have been identified as driving forces to stabilize the self-assembled structures of metallopolymers [8, 25, 26]. Studies also showed that most metallopolymers were under thermodynamically changing processes with change of temperature [8, 12, 25]. Ian Manners group synthesized various self-assembled metallopolymers, and studied its formation mechanism and its applications in nanolithography, biomedicine, magnetic or responsive materials (**Figure 2**) [24]. Till now, many supramolecular metallopolymers of gold, copper and platinum complexes, have been well documented [2, 10, 17, 24, 26–30], and the search for application potential has become new research focus.

While most of supramolecular metallopolymers are "high-molecular-weight" with relatively large ligands [31], the discovery of low molecular-weight metallopolymers with tunable structures have fostered a new growth in recent years [3, 4, 32–35]. Rissanen et al. reported a terpyridine-Zn (II) compound that can self-assemble into metallopolymers with fibrous structured microscopic morphology. Their studies show that this compound can realize the detection of nanomole pyrophosphate in aqueous solution and the detection of pyrophosphate in the competitive environment of cytoplasmic ions [33]. Study has showed a low-molecular-weight metallopolymer in nanofiber form demonstrating potent anticancer properties. [17, 36] Nano-formulation has been showed to be able to improve the stability and selectivity of metal complexes, and hence emerges as an appealing strategy to increase anticancer activity and reduce their toxic side effects [37–39]. For example, Wu et al. have synthesized photosensitive triblock copolymers, which can be self-assembled to form polymer nanosystems to improve their biocompatibility and prolong their blood circulation time to achieve the purpose of regional enrichment of tumor tissue. Under the excitation of red light, the nanosystem dissociates spontaneously, releases anticancer Ru complexes and triggers the generation of a large amount of singlet oxygen to inhibit tumor growth [37]. Therefore, the formation of nanostructures of metallopolymer could have a

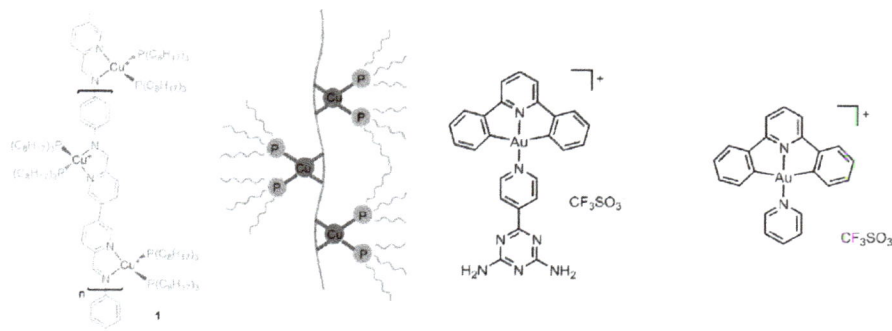

Figure 1.
Schematic representation of polymer and chemical structure of [Au^III(C^N^C)(4-dpt)]-(CF₃SO₃) from Ref. [12, 17].

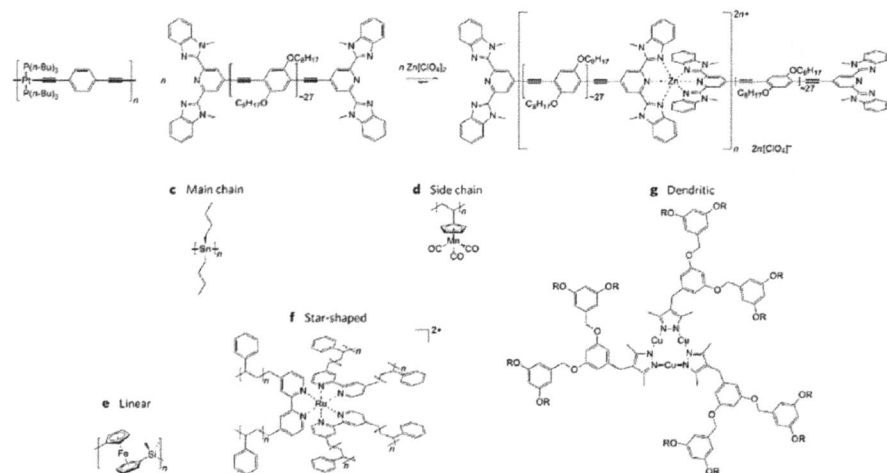

Figure 2.
Metallopolymers of different metal centers [24].

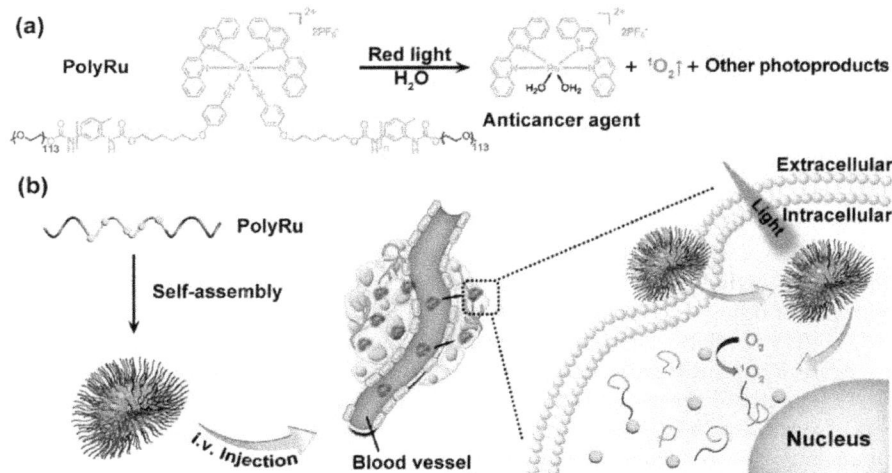

Figure 3.
(a) Structure and photo-induced dissociation of Ru complexes, (b) synthesis of nano-system and its inhibitory effect on tumor growth [37].

promising improvement on the biological activities. Recently, studies have found that, metal complexes with 2-phenylimidazo [4,5-f]-[1,10]phenanthroline (pip) as ligand displayed potent anticancer activities (**Figure 3**) [17, 34, 40–43].

2. Synthesis and characterization

Possibly, the plane structure and NH group of the ligand could form π-π interactions and hydrogen bonds between adjacent molecules. Interestingly, in this study, we synthesized a simple Cu(II) complex, [CuCl(pip)₂]Cl, capable of self-assembling into a metallopolymer driven by diverse intermolecular interaction,

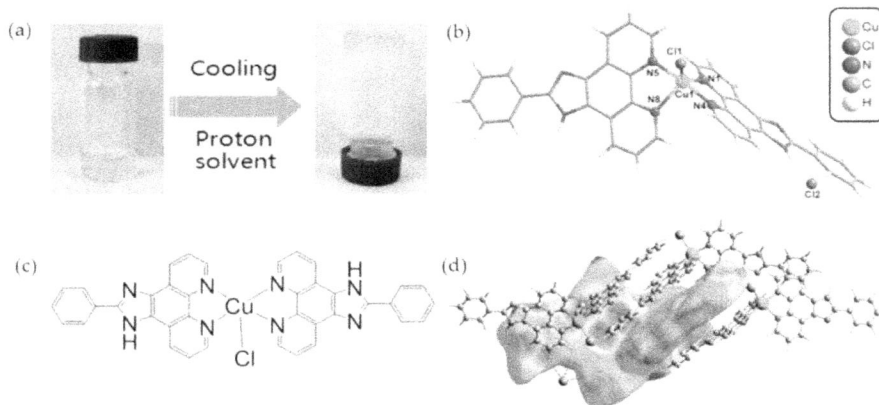

Figure 4.
[CuCl(pip)$_2$]Cl: (a) formation of the viscous fluid upon cooling in ethanol/water (v/v = 5:1). (b) Crystal structure, (c) chemical structure, (d) illustration of Hirshfeld surface in the crystal packing: the mapping range is shown from red (short distance) through green to blue (long distance) [44].

which demonstrated potent in vivo anticancer efficacy. Addition of pip in ethanol to CuCl$_2$ in water resulted in a clear green solution. The solution turned to be viscous after 2 h at room temperature. Upon cooling in a refrigerator, stable green viscous solution was formed (**Figure 4a**). When the viscous solution stood at room temperature for 25 days, green single crystals suitable for X-ray diffraction could be obtained in a capped vial. The crystal unit of [CuCl(pip)$_2$]Cl is in a trigonal pyramidal coordination geometry (**Figure 4b**). Hirshfeld surface [45] clearly illustrates that rich supramolecular interactions of conjugated π-π interactions, C(N)−H × × × Cl hydrogen bonds, and edge-to-face C−H × × × π interactions are involved (**Figure 4c** and **d**) in the crystal packing. These results indicate that weak interactions including cooperative π-π and multiple unconventional C−H···X hydrogen bonding interactions are strong enough to from the metallopolymer.

UV/Vis and FT-IR spectroscopy was employed to examine the intermolecular aggregation of [CuCl(pip)$_2$]Cl during the assembly process. As shown in **Figure 5a**, UV/Vis spectra of the complex in CH$_3$CH$_2$OH/H$_2$O (5:1) exhibits a red shift with peaks from 409 to 423 nm when the concentration increased. This is ascribed to the enhancement of molecular interactions due to the increase of the concentration. The FT-IR spectra of [CuCl(pip)$_2$]Cl displayed significant difference in different solvents (**Figure 5b**). The red shifts in proton solvent could be attributed to the enhancement of C−H × × × X hydrogen bond and π-π interaction.

Tyndall effect was observed in the solution of [CuCl(pip)$_2$]Cl in proton solvent CH$_3$CH$_2$OH/H$_2$O (**Figure 6**), indicating the formation of self-assembled species [46], while no Tyndall phenomenon observed in aprotic solvent DMF. Therefore, the presence of proton solvent could be a determining factor for the self-assembly, as proton solvent could provide H atom feasible for the aggregation driven by hydrogen bond. In contrast, Tyndall phenomenon could not be observed in the solution of 0.8 mM (**Figure 6a**), possibly due to the unfavorable intermolecular distances in a dilute solution. Moreover, the mean size of the metallopolymer was found at about 95 nm (**Figure 6b**), indicating the presence of nanoscale aggregates in this solution.

From the TEM images, we found that metallopolymer were highly monodisperse with the size of 83 nm in diameter. And AFM measurement confirmed the spherical nanoparticle morphology of the supramolecular metallopolymer in the

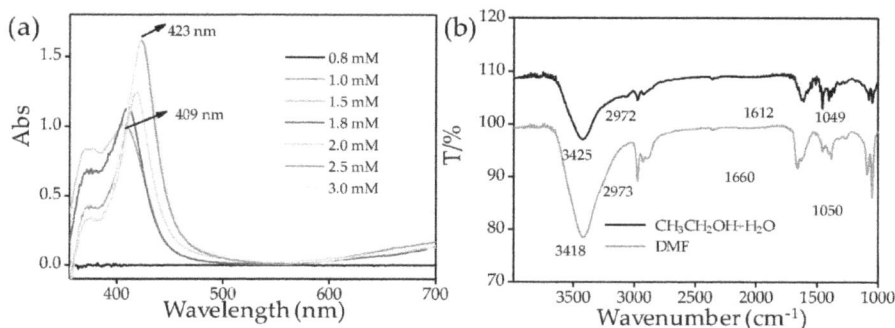

Figure 5.
UV/Vis (a) and FT-IR (b) spectroscopy changes in the self-assembly process [44].

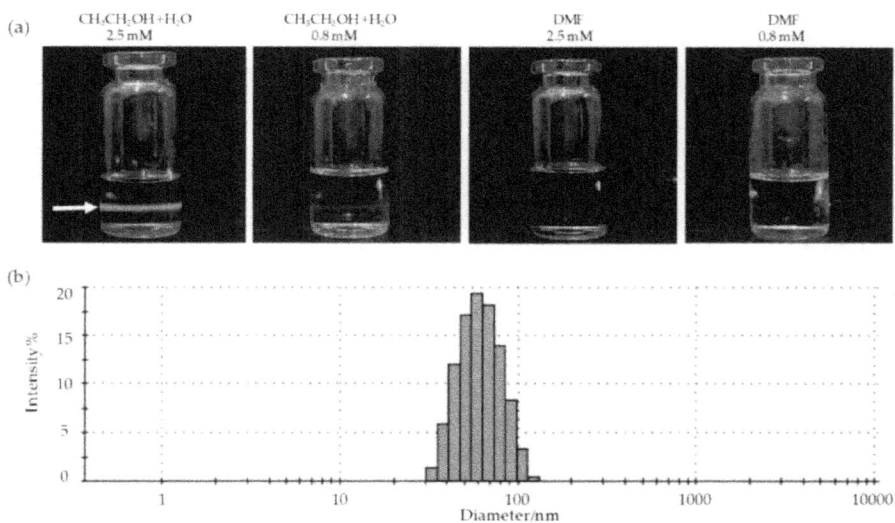

Figure 6.
Nanostructure of metallopolymer. (a) Tyndall effect of [CuCl(pip)$_2$]Cl under different solvent and concentration conditions. (b) Size distribution of [CuCl(pip)$_2$]Cl (2.5 mM) in ethanol/water (5:1) [44].

proton solvent (**Figure 7a** and **b**). The metallopolymer was further confirmed by XRD (**Figure 7c**). The results of MALDI-TOF-MS analysis also demonstrated the presence of monomeric species, and dimer and trimer aggregation peaks of [CuCl(pip)$_2$]Cl (**Figure 7d**).

To further understand the phase transition, the effects of temperature, concentration and solvent on the viscosity of [CuCl(pip)$_2$]Cl were examined by Ubbelohde viscometer. Significant temperature-, solvent- and concentration-dependent changes in the viscosity were recorded (**Figure 8b** and **c**). The stacking mode of [CuCl(pip)$_2$]Cl showing hydrogen bonds (C–H...Cl = 3.681 Å; N–H···Cl = 3.279 Å) and π-π interactions (3.524–3.777 Å) (**Figure 8a**). High viscosity was detected when incubating in CH$_3$CH$_2$OH/H$_2$O (5:1), which decreased obviously when the temperature increased from 5 to 75°C. However, the viscosity showed no significant increase in aprotic solvent DMF (**Figure 8d**). Additionally, the decreased viscosity induced by rising temperature when the temperature decreased from 75 to 5°C (**Figure 8e**), which demonstrated the recovering ability of the metallopolymer. The viscosity of

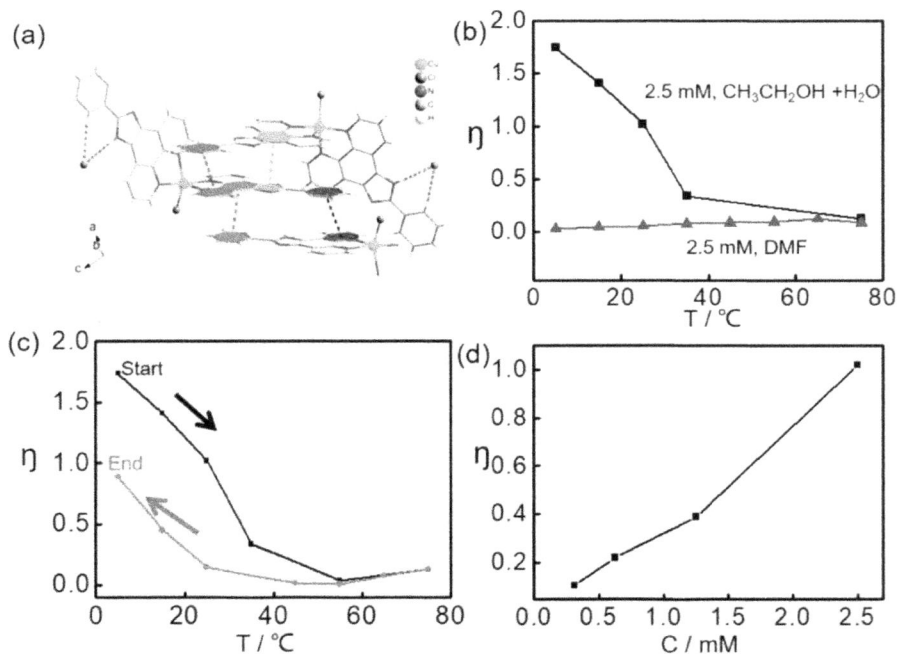

Figure 7.
Characterization of metallopolymer. TEM (a) and AFM (b) images of metallopolymer. (c) Powder XRD pattern of [CuCl(pip)₂]Cl (2.5 mM) in ethanol/water (5:1). (d) MALDI-TOF-MS analysis of the [CuCl(pip)₂]Cl in ethanol/water (5:1). Insets specific peaks representing different aggregation patterns [44].

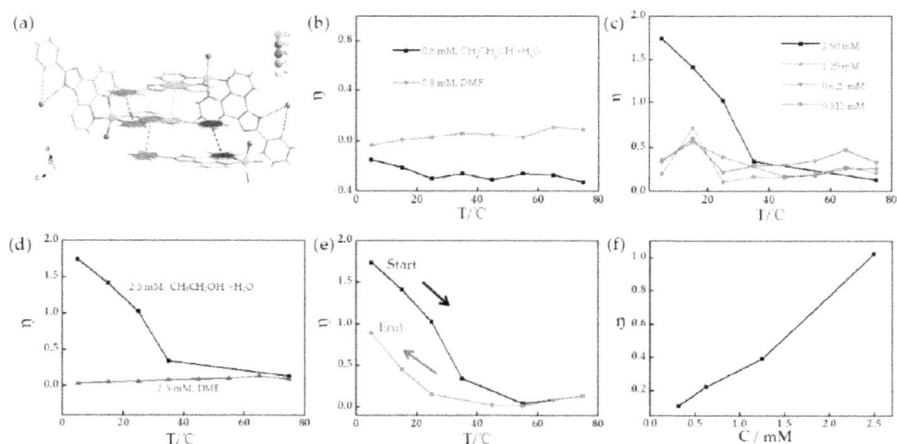

Figure 8.
Dynamic change of metallopolymer with different treatment. (a) The stacking mode of [CuCl(pip)₂]Cl showing hydrogen bonds (C–H...Cl = 3.681 Å; N–H···Cl = 3.279 Å) and π-π interactions (3.524–3.777 Å), and viscosity change of [CuCl(pip)₂]Cl in different solvents and temperatures, (b) viscosity of [CuCl(pip)₂] Cl (0.8 mM) dissolved in different solvents with temperature range from 5 to 75°C, (c) the viscosity of [CuCl(pip)₂]Cl was recorded on temperature variation with the concentration raised, (d) dynamic change with temperature varying in the cycle 5–75–5°C in ethanol/water (5:1), (e) concentration dependent at 25°C, and (f) dynamic change of metallopolymer with different concentration [44].

[CuCl(pip)$_2$]Cl increased dramatically upon increasing the concentration, which confirms the contribution of enhanced intermolecular interactions during the self-assembly of the metallopolymer (**Figure 8f**).

3. Cell proliferation inhibition of metallopolymer

Further studies were also carried out to examine the effects of the matter forms (metallopolymer and monomeric complex) on the anticancer activity of [CuCl(pip)$_2$]Cl, by in vitro and in vivo models as previously described [39, 47]. As shown in **Figure 9a**, the metallopolymer demonstrated much higher anticancer activities against the tested cancer cells than the monomeric complex. In order to understand the reasons accounting for the different activities induced by the matter forms, we compared their cellular uptake in HepG2 hepatocellular carcinoma cells. Consistently, the metallopolymer exhibited much higher cellular uptake than monomeric complex in different time points (**Figure 9b**). From **Figure 9c**, we found that significant concentration-dependent changes in the cellular uptake were

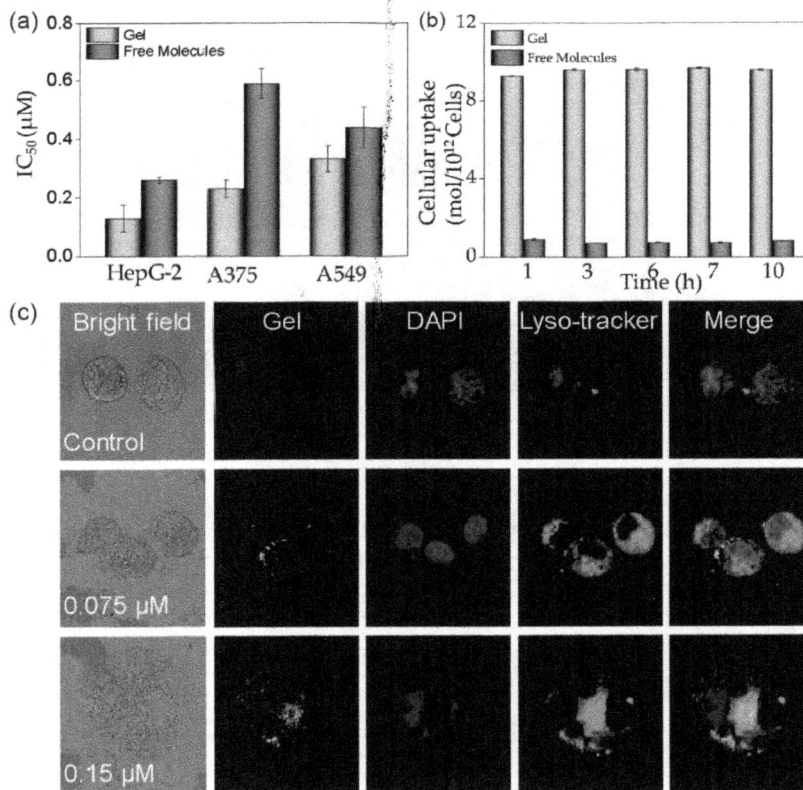

Figure 9.
Cancer cell growth inhibition of metallopolymer and monomeric [CuCl(pip)$_2$]Cl. (a) The cytotoxic effects of monomeric [CuCl(pip)$_2$]Cl, metallopolymer, (b) time-course cellular uptake of monomeric [CuCl(pip)$_2$]Cl, metallopolymer in HepG2 cells (10 μM), and (c) cellular uptake of metallopolymer by confocal fluorescence images [44].

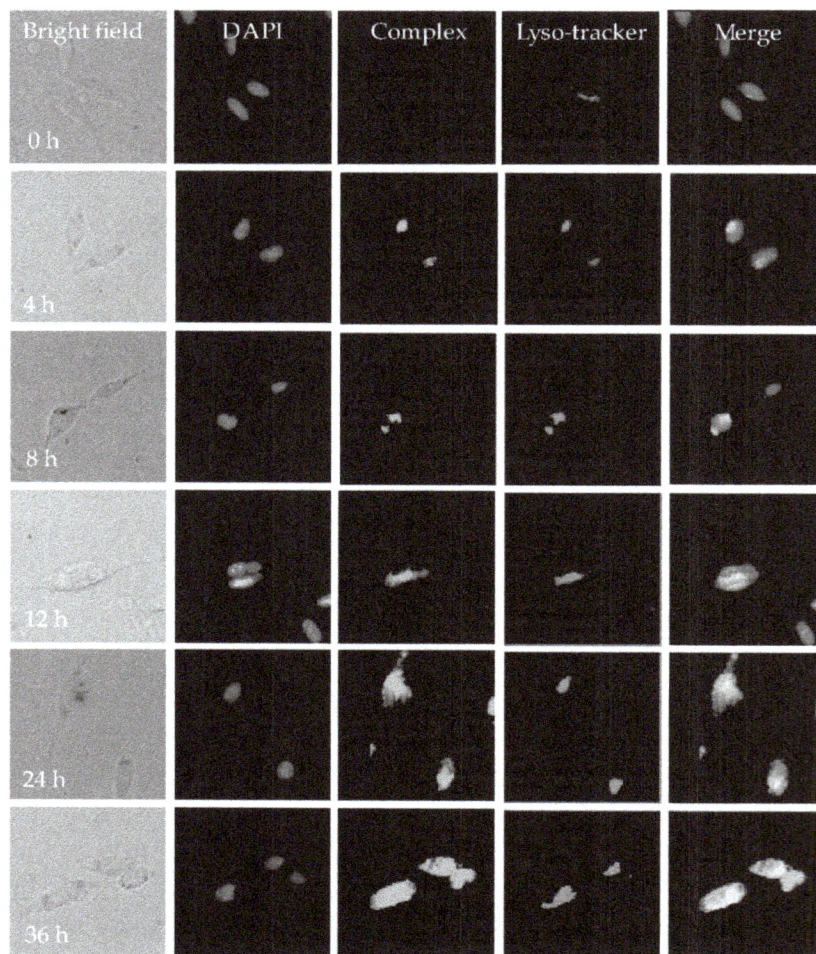

Figure 10.
Cellular localization of metallopolymer after incubation with HepG2 cells for different periods of time [44].

recorded. It is likely that metallopolymer could assembly into nanoparticles, which enter cancer cells through endocytosis in a high efficiency, thus increasing the cellular uptake and anticancer efficacy. Moreover, the results of confocal fluorescence images revealed that, the metallopolymer were internalized by cancer cells through endocytosis, and could be released into cytoplasm after 12–24 h (**Figure 10**).

4. In vivo tumor growth inhibition of metallopolymer

Furthermore, we assessed the in vivo therapeutic efficacy of the metallopolymer in HepG2 xenografts nude mice. In this study, metallopolymer was dispersed in PBS and injected into the tumor-bearing nude mice intravenously. As shown in **Figure 11**, the metallopolymer significantly inhibited the tumor growth, as evidenced by the decrease in tumor volume and tumor weight in a time-dependent manner. Moreover, under the effective dose, the metallopolymer showed no damage to these major organs, including heart, liver, spleen, lung and kidney (**Figure 12**),

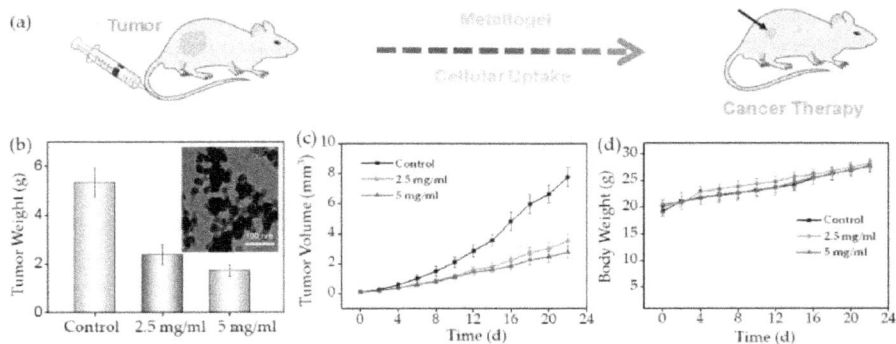

Figure 11.
(a) Schematic demonstration for tumor growth inhibition by metallopolymer. (b and c) In vivo anticancer activity of the metallopolymer against HepG2 cells xenografts. Inset is TEM images of the metallopolymer in DMEM medium (24 h), and (d) body weight of HepG2 cells xenografts in nude mice (n = 5) [44].

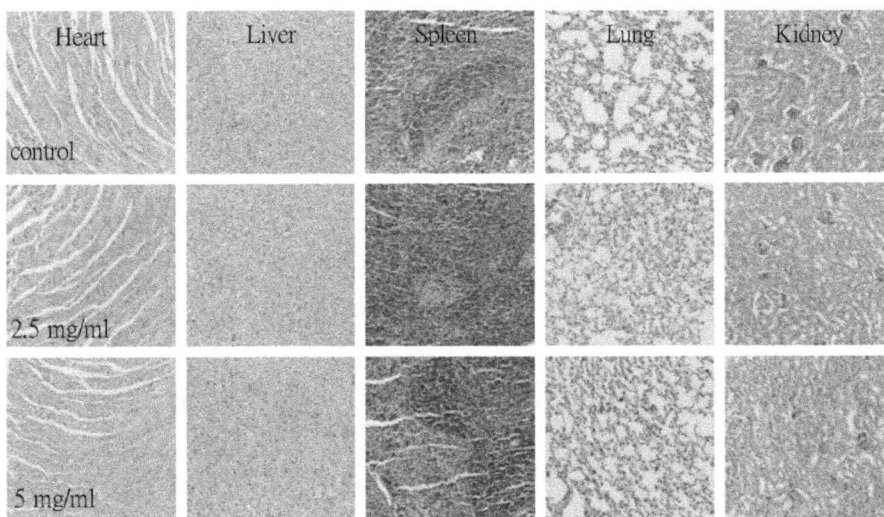

Figure 12.
H&E staining of major organs after treatments [44].

demonstrating the cancer therapeutic potential and safety of this kind of self-assembled functional metallopolymer.

5. Conclusions and prospects

In the past decade, self-assembled functional supramolecular metallopolymers have aroused a surge of research interest, and have demonstrated application potential in cancer therapy. In this chapter, we have summarized the progress in the rational design of biological application of different metallopolymers. Especially, a simple Cu(II) complex, [CuCl(pip)$_2$]Cl, was found be able to self-assemble into surpramolecular metallopolymer driven by diverse intermolecular interactions, including π-π interactions and hydrogen bonds under a proton solvent condition. The functional metallopolymer could enter cancer cells through endocytosis, thus effectively inhibit tumor growth in vivo without damage to the major organs. This study provides a simple strategy for rational design of Cu-based metallopolymer with novel anticancer potency. For further development of self-assembled nanostructures with clinical application prospect and value in tumor diagnosis and treatment, here are some issues that should be considered. (1) Firstly, how to make the self-assembly of compounds more efficient? This may require structural optimization of ligands, metal centers, bonding methods, or modification of polymers or biological macromolecules to adjust the overall hydrophilicity and lipophilicity of complex molecules. (2) Secondly, how to make the assembly more selective and targeted to the tumor? It is necessary to transform the structure or add targeted groups to make the assembly more specific to tumor tissue, cells and intracellular molecules, so as to distinguish normal cells. (3) In order to use as clinic treatment, more research is needed in the aspect of medicinal properties, such as toxicology, pharmacokinetic analysis, etc., so that researchers can have a better understanding of the medicinal properties of different types of assembled structures.

Acknowledgements

This work was supported by National Natural Science Foundation of China (21877049), National Program for Support of Top-notch Young Professionals (W02070191), YangFan Innovative & Entrepreneurial Research Team Project (201312H05) and Fundamental Research Funds for the Central Universities.

Conflict of interest

There are no conflicts to declare.

Author details

Zushuang Xiong, Lanhai Lai and Tianfeng Chen*
Department of Chemistry, Jinan University, Guangzhou, China

*Address all correspondence to: felixchentf@gmail.com

IntechOpen

References

[1] Winter A, Schubert US. Synthesis and characterization of metallo-supramolecular polymers. Chemical Society Reviews. 2016;**45**(19):5311-5357. DOI: 10.1039/C6CS00182C

[2] Theis S, Iturmendi A, Gorsche C, et al. Metallo-supramolecular gels that are photocleavable with visible and near-infrared irradiation. Angewandte Chemie International Edition. 2017;**56**(50):15857-15860. DOI: 10.1002/anie.201707321

[3] Yang L, Tan X, Wang Z, et al. Supramolecular polymers: Historical development, preparation, characterization, and functions. Chemical Reviews. 2015;**115**(15): 7196-7239. DOI: 10.1021/cr500633b

[4] Wu H, Zheng J, Kjøniksen AL, et al. Metallogels: Availability, applicability, and advanceability. Advanced Materials. 2019;**31**(12):1806204. DOI: 10.1002/adma.201806204

[5] Wei P, Yan X, Huang F. Supramolecular polymers constructed by orthogonal self-assembly based on host-guest and metal-ligand interactions. Chemical Society Reviews. 2015;**44**(3):815-832. DOI: 10.1039/C4CS00327F

[6] Lin Q, Lu TT, Zhu X, et al. A novel supramolecular metallogel-based high-resolution anion sensor array. Chemical Communications. 2015;**51**(9):1635-1638. DOI: 10.1039/C4CC07814D

[7] Dong R, Zhou Y, Huang X, et al. Functional supramolecular polymers for biomedical applications. Advanced Materials. 2015;**27**(3):498-526. DOI: 10.1002/adma.201402975

[8] Janeček ER, McKee JR, Tan CSY, et al. Hybrid supramolecular and colloidal hydrogels that bridge multiple length scales. Angewandte Chemie International Edition. 2015;**54**(18): 5383-5388. DOI: 10.1002/anie.201410570

[9] Goor OJGM, Hendrikse SIS, Dankers PYW, et al. From supramolecular polymers to multi-component biomaterials. Chemical Society Reviews. 2017;**46**(21):6621-6637. DOI: 10.1039/C7CS00564D

[10] Pal A, Malakoutikhah M, Leonetti G, et al. Controlling the structure and length of self-synthesizing supramolecular polymers through nucleated growth and disassembly. Angewandte Chemie International Edition. 2015;**54**(27): 7852-7856. DOI: 10.1002/anie.201501965

[11] Lahav M, van der Boom ME. Polypyridyl metallo-organic assemblies for electrochromic applications. Advanced Materials. 2018;**30**(41):1706641. DOI: 10.1002/adma.201706641

[12] de Hatten X, Bell N, Yufa N, et al. A dynamic covalent, luminescent metallopolymer that undergoes sol-to-gel transition on temperature rise. Journal of the American Chemical Society. 2011;**133**(9):3158-3164. DOI: 10.1021/ja110575s

[13] Zhou Y, Zhang HY, Zhang ZY, et al. Tunable luminescent lanthanide supramolecular assembly based on photoreaction of anthracene. Journal of the American Chemical Society. 2017;**139**(21):7168-7171. DOI: 10.1021/jacs.7b03153

[14] Gao Z, Han Y, Wang F. Cooperative supramolecular polymers with anthracene-endoperoxide photo-switching for fluorescent anti-counterfeiting. Nature Communications. 2018;**9**(1):1-9. DOI: 10.1038/s41467-018-06392-x

[15] Li J, Su Z, Xu H, et al. Supramolecular networks of hyperbranched poly (ether amine) (hPEA) nanogel/chitosan (CS) for the selective adsorption and separation of guest molecules. Macromolecules. 2015;**48**(7):2022-2029. DOI: 10.1021/ma502607p

[16] Li H, Sadiq MM, Suzuki K, et al. Magnetic metal-organic frameworks for efficient carbon dioxide capture and remote trigger release. Advanced Materials. 2016;**28**(9):1839-1844. DOI: 10.1002/adma.201505320

[17] Zhang JJ, Lu W, Sun RWY, et al. Organogold (III) supramolecular polymers for anticancer treatment. Angewandte Chemie International Edition. 2012;**51**(20):4882-4886. DOI: 10.1002/anie.201108466

[18] Mahadevi AS, Sastry GN. Cooperativity in noncovalent interactions. Chemical Reviews. 2016;**116**(5):2775-2825. DOI: 10.1021/cr500344e

[19] Kuosmanen R, Rissanen K, Sievänen E. Steroidal supramolecular metallogels. Chemical Society Reviews. 2020. DOI: 10.1039/C9CS00686A

[20] Kurbah SD, Lal RA. Vanadium (V) complex based supramolecular metallogel: Self-assembly and (metallo) gelation triggered by non-covalent and N⁺H...O hydrogen bonding interactions. Inorganic Chemistry Communications. 2020;**111**:107642. DOI: 10.1016/j.inoche.2019.107642

[21] Fu HLK, Leung SYL, Yam VWW. A rational molecular design of triazine-containing alkynylplatinum (ii) terpyridine complexes and the formation of helical ribbons via Pt···Pt, π-π stacking and hydrophobic-hydrophobic interactions. Chemical Communications. 2017;**53**(82):11349-11352. DOI: 10.1039/C7CC06293A

[22] Zhou Z, Yan X, Cook TR, et al. Engineering functionalization in a supramolecular polymer: Hierarchical self-organization of triply orthogonal non-covalent interactions on a supramolecular coordination complex platform. Journal of the American Chemical Society. 2016;**138**(3):806-809. DOI: 10.1021/jacs.5b12986

[23] Okesola BO, Smith DK. Applying low-molecular weight supramolecular gelators in an environmental setting—Self-assembled gels as smart materials for pollutant removal. Chemical Society Reviews. 2016;**45**(15):4226-4251. DOI: 10.1039/C6CS00124F

[24] Whittell GR, Hager MD, Schubert US, et al. Functional soft materials from metallopolymers and metallosupramolecular polymers. Nature Materials. 2011;**10**(3):176-188. DOI: 10.1038/nmat2966

[25] Chan MHY, Ng M, Leung SYL, et al. Synthesis of luminescent platinum (II) 2, 6-bis (N-dodecylbenzimidazol-2′-yl) pyridine foldamers and their supramolecular assembly and metallogel formation. Journal of the American Chemical Society. 2017;**139**(25):8639-8645. DOI: 10.1021/jacs.7b03635

[26] Fu HLK, Po C, He H, et al. Tuning of supramolecular architectures of l-valine-containing dicyanoplatinum (II) 2, 2′-bipyridine complexes by metal-metal, π-π stacking, and hydrogen-bonding interactions. Chemistry - A European Journal. 2016;**22**(33):11826-11836. DOI: 10.1002/chem.201601983

[27] Wang Y, Astruc D, Abd-El-Aziz AS. Metallopolymers for advanced sustainable applications. Chemical Society Reviews. 2019;**48**(2):558-636. DOI: 10.1039/C7CS00656J

[28] Shi B, Zhou Z, Vanderlinden RT, et al. Spontaneous supramolecular

polymerization driven by discrete platinum metallacycle-based host-guest complexation. Journal of the American Chemical Society. 2019;**141**(30): 11837-11841. DOI: 10.1021/jacs.9b06181

[29] Kumar A, Bawa S, Ganorkar K, et al. Synthesis and characterization of acid-responsive luminescent Fe (II) metallopolymers of rigid and flexible backbone N-donor multidentate conjugated ligands. Inorganic Chemistry. 2020. DOI: 10.1021/acs. inorgchem.9b02985

[30] Gee WJ, Batten SR. Instantaneous gelation of a new copper (II) metallogel amenable to encapsulation of a luminescent lanthanide cluster. Chemical Communications. 2012;**48**(40):4830-4832. DOI: 10.1039/ C2CC30170A

[31] Zhang KY, Liu S, Zhao Q, et al. Stimuli-responsive metallopolymers. Coordination Chemistry Reviews. 2016;**319**:180-195. DOI: 10.1016/j. ccr.2016.03.016

[32] Kuwahara RY, Yamagishi H, Hashimoto K, et al. Easy preparation and characterization of conducting polymer-low molecular weight organogel system. Polymer. 2015;**61**:99-107. DOI: 10.1016/j. polymer.2014.12.059

[33] Bhowmik S, Ghosh BN, Marjomäki V, et al. Nanomolar pyrophosphate detection in water and in a self-assembled hydrogel of a simple terpyridine-Zn^{2+} complex. Journal of the American Chemical Society. 2014;**136**(15):5543-5546. DOI: 10.1021/ ja4128949

[34] Bentz KC, Cohen SM. Supramolecular metallopolymers: From linear materials to infinite networks. Angewandte Chemie International Edition. 2018;**57**(46):14992-15001. DOI: 10.1002/anie.201806912

[35] Yan D, Evans DG. Molecular crystalline materials with tunable luminescent properties: From polymorphs to multi-component solids. Materials Horizons. 2014;**1**(1):46-57. DOI: 10.1039/C3MH00023K

[36] Tsai JLL, Zou T, Liu J, et al. Luminescent platinum (II) complexes with self-assembly and anti-cancer properties: Hydrogel, pH dependent emission color and sustained-release properties under physiological conditions. Chemical Science. 2015;**6**(7):3823-3830. DOI: 10.1039/ C4SC03635B

[37] Sun W, Li S, Häupler B, et al. An amphiphilic ruthenium polymetallodrug for combined photodynamic therapy and photochemotherapy in vivo. Advanced Materials. 2017;**29**(6):1603702. DOI: 10.1002/ adma.201603702

[38] Lai H, Zhang X, Feng P, et al. Enhancement of antiangiogenic efficacy of iron (II) complex by selenium substitution. Chemistry, an Asian Journal. 2017;**12**(9):982-987. DOI: 10.1002/asia.201700272

[39] Zhao Z, Zhang X, Li C, et al. Designing luminescent ruthenium prodrug for precise cancer therapy and rapid clinical diagnosis. Biomaterials. 2019;**192**:579-589. DOI: 10.1016/j. biomaterials.2018.12.002

[40] Lai H, Zeng D, Liu C, et al. Selenium-containing ruthenium complex synergizes with natural killer cells to enhance immunotherapy against prostate cancer via activating TRAIL/FasL signaling. Biomaterials. 2019;**219**:119377. DOI: 10.1016/j. biomaterials.2019.119377

[41] Zhao J, Zhang X, Liu H, et al. Ruthenium arene complex induces cell cycle arrest and apoptosis through activation of P53-mediated signaling pathways. Journal of Organometallic

Chemistry. 2019;**898**:120869. DOI: 10.1016/j.jorganchem.2019.07.020

[42] Xie L, Luo Z, Zhao Z, et al. Anticancer and antiangiogenic iron (II) complexes that target thioredoxin reductase to trigger cancer cell apoptosis. Journal of Medicinal Chemistry. 2017;**60**(1):202-214. DOI: 10.1021/acs.jmedchem.6b00917

[43] Zhao Z, Gao P, You Y, et al. Cancer-targeting functionalization of selenium-containing ruthenium conjugate with tumor microenvironment-responsive property to enhance theranostic effects. Chemistry - A European Journal. 2018;**24**(13):3289-3298. DOI: 10.1002/chem.201705561

[44] Lai L, Luo D, Liu T, et al. Self-assembly of copper polypyridyl supramolecular metallopolymers to achieve enhanced anticancer efficacy. ChemistryOpen. 2019;**8**:434-443. DOI: 10.1002/open.201900036

[45] Seth SK, Bauzá A, Frontera A. Screening polymorphism in a Ni (II) metal-organic framework: Experimental observations, Hirshfeld surface analyses and DFT studies. CrystEngComm. 2018;**20**(6):746-754. DOI: 10.1039/C7CE01991B

[46] Xiao W, Deng Z, Huang J, et al. Highly sensitive colorimetric detection of a variety of analytes via the Tyndall effect. Analytical Chemistry. 2019;**91**(23):15114-15122. DOI: 10.1021/acs.analchem.9b03824

[47] Deng Z, Gao P, Yu L, et al. Ruthenium complexes with phenylterpyridine derivatives target cell membrane and trigger death receptors-mediated apoptosis in cancer cells. Biomaterials. 2017;**129**:111-126. DOI: 10.1016/j.biomaterials.2017.03.017